MW00768028

PORT FACILITIES AND CONTAINER HANDLING SERVICES

Leonard K. Cheng

Yue-Chim Richard Wong

Published for

The Hong Kong Centre for Economic Research

The Better Hong Kong Foundation

The Hong Kong Economic Policy Studies Forum

By

City University of Hong Kong Press

First published 1997
Printed in Hong Kong

ISBN 962-937-003-4

Published by
City University of Hong Kong Press
City University of Hong Kong
Tat Chee Avenue, Kowloon, Hong Kong

Internet: http://www.cityu.edu.hk/upress/
E-mail: upress@cityu.edu.hk

The free-style calligraphy on the cover, *gang,* means "harbour" in Chinese.

Contents

Detailed Chapter Contents

Foreword

The key to the economic success of Hong Kong has been a business and policy environment which is simple, predictable and transparent. Experience shows that prosperity results from policies that protect private property rights, maintain open and competitive markets, and limit the role of the government.

The rapid structural change of Hong Kong's economy in recent years has generated considerable debate over the proper role of economic policy in the future. The impending restoration of sovereignty over Hong Kong from Britain to China has further complicated the debate. Anxiety persists as to whether the pre-1997 business and policy environment of Hong Kong will continue.

During this period of economic and political transition in Hong Kong, various interested parties will be re-assessing Hong Kong's existing economic policies. Inevitably, some will advocate an agenda aimed at altering the present policy making framework to reshape the future course of public policy.

For this reason, it is of paramount importance for those familiar with economic affairs to reiterate the reasons behind the success of the economic system in the past, to identify what the challenges are for the future, to analyze and understand the economy sector by sector, and to develop appropriate policy solutions to achieve continued prosperity.

In a conversation with my colleague Y. F. Luk, we came upon the idea of inviting economists from universities in Hong Kong to take up the challenge of examining systematically the economic policy issues of Hong Kong. An expanding group of economists (The Hong Kong Economic Policy Studies Forum) met several times to give form and shape to our initial ideas. The Hong Kong Economic Policy Studies Project was then launched in 1996 with some 30 economists from the universities in Hong Kong and a few from overseas. This is the first time in Hong Kong history that a concerted public effort has been undertaken by academic economists in the territory. It represents a joint expression of our collective concerns, our hopes for a better Hong Kong, and our faith in the economic future.

The Hong Kong Centre for Economic Research is privileged to be co-ordinating this Project. We are particularly grateful to The Better Hong Kong Foundation whose support and assistance has made it possible for us to conduct the present study, the results of which are published in this monograph. We also thank the directors and editors of the City University of Hong Kong Press and The Commercial Press (H.K.) Ltd. for their enthusiasm and dedication which extends far beyond the call of duty. The unfailing support of many distinguished citizens in our endeavour and their words of encouragement are especially gratifying.

Yue-Chim Richard Wong
Director
The Hong Kong Centre
for Economic Research

Foreword by Series Editor

Port facilities and container handling services may be of little direct relevance to the daily economic life of most people in Hong Kong. Yet the significance of these two services to the whole economy should not be underestimated.

Hong Kong had become a major regional entrepôt before the Second World War broke out. After a lapse of some 30 years, Hong Kong regained its leading position in shipping by the early 1980s. The volume of freight moving through Hong Kong multiplied rapidly in the past decade and a half. This has been obviously the result of Mainland China's re-entrance to the world market, and of the growth and development in the rest of the region. The good geographical location and the economic flexibility of Hong Kong have enabled it to take advantage of the booming trade in the region.

Hong Kong has had many years of experience in port development, especially in handling container cargoes. Now is an appropriate time to assess Hong Kong's experience and its prospects in maintaining the position as the world's busiest container port. There are two broad categories of relevant issues to analyze.

The first category of issues are mainly domestic. Questions concerning the cargoes are: What has been the changing importance of different types of cargoes, the different modes of cargo handling services, and the different geographical sources and destinations of the cargo flows? How have these changes affected Hong Kong as a port and its cargo handling industry? Given the rising importance of containerized cargoes, will we have enough facilities to accommodate the forecasted demand for container handling services? As to questions concerning the industry. How is the industry organized at present? Is the industrial structure conducive to economic efficiency? Is there sufficient competition in container handling services? These are all important issues for the planning and future performance of Hong Kong as a major cargo hub of the East and Southeast Asia Region.

The second category of issues are external in nature. Economic growth and expanding global trade have led to the development of new port facilities in Mainland China. Given the large share of Mainland China in total shipment through Hong Kong, the parallel developments of ports in China is posing new challenges to the leadership of Hong Kong in regard to cargo handling. Similar port developments are also taking place elsewhere in the Pacific region. Will these emerging regional ports necessarily become rivalries? Can they be complementary? Can Hong Kong maintain its leading edge in this network of ports in the future? What are the relative strengths and weaknesses of Hong Kong vis-à-vis its competitors?

This book discusses all the above issues in detail. The authors are perhaps the first in Hong Kong to have comprehensively analysed the port facilities and container handling services of the territory. There have been sporadic writings on this topic, but this book is certainly much more extensive in coverage and analytical in methodology.

The authors, Professors Leonard Kwok-hon Cheng and Richard Yue-chim Wong, examine the issues in wider economic perspectives, especially those of international trade and industrial organization. They also make forecasts of the future demand for container handling services in Hong Kong. The outcome is a monograph that should interest not only those practitioners and policy makers in the field, but also some economic analysts and the general public who are concerned with the future of Hong Kong as a trade centre. They will all find this work helpful. This book will be the standard reference in related discussions in the future, especially in regard to the controversial public-policy issues regarding competitive strategies in major infrastructural investments.

Y. F. Luk
School of Economics and Finance
The University of Hong Kong

Preface

During our research we have benefited greatly from the assistance of these organizations and individuals: Patrick P. W. Chan and Henry P. S. Wah of Shanghai Container Terminals; Wing Kee Chan and Clement Yeung of Hong Kong Shippers' Council; Philip Chow, C. H. Tung, and Allan T. S. Wong of Orient Overseas Container Line; Tony Clark of the Port Development Board; John Harries, Richard Pearson, and James S. Tsien of Hongkong International Terminals; Stanley K. C. Ko of Jardine Pacific; Victor Li of Cheung Kong; Lu Bohong of Ningpo Port Authority; Tony Miller of the Department of Trade; Wilson Kong and Dennis Ng of Sea-Land Orient Terminals; Michael B. Sandpearl of Modern Terminals Limited; Captain Ngai Lap Chee of Marine Investigation and Consultants Ltd.; Chi Schive of Economic Development Committee, Taiwan; Kenneth K. T. Tse of Yantian International Container Terminals; and Mak Nak Keung of Sun Hung Kai Properties.

We would like to express our gratitude to the above individuals and organizations for their input and assistance and to thank our research assistance Winnie Lam, Kenneth Ma, and Amy Wong. They do not necessarily share our views and positions. Finally, we are indebted to Virginia Unkefer for her editorial assistance.

Leonard K. Cheng
The Hong Kong University of Science and Technology
Yue-Chim Richard Wong
The University of Hong Kong

List of Illustrations

Figures

Tables

Acronyms and Abbreviations

Acronyms	Names in Full	Appear First on Page
ACT	Asia Container Terminals	26
ASEAN	Association of Southeast Asian Nations	10
COSCO–HIT	China Ocean Shipping Company – Hongkong International Terminals	24
FEU	Forty Foot Equivalent Unit	50
GDP	Gross Domestic Product	83
GNP	Gross National Product	83
HIT	Hongkong International Terminals	3
HWD	Hutchison Whampoa Dock Company	24
IMF	International Monetary Fund	87
MOFTEC	Ministry of Foreign Trade and Economic Cooperation	21
MTL	Modern Terminals Limited	3
OCL	Overseas Container Line	23
OECD	Organisation of Economic Cooperation and Development	87
OOCL	Orient Overseas Container Line	24
OPEC	Organization of Petroleum Exporting Countries	57
PDB	Port Development Board	2
PRC	People's Republic of China	64
SCT	Shanghai Container Terminals	69
SLOT	Sea-Land Orient Terminals	24
TEU	Twenty Foot Equivalent Unit	16
THCs	Terminal Handling Charges	44
WTO	World Trade Organization	1
YICT	Yantian International Container Terminals	63

Ports in the Region

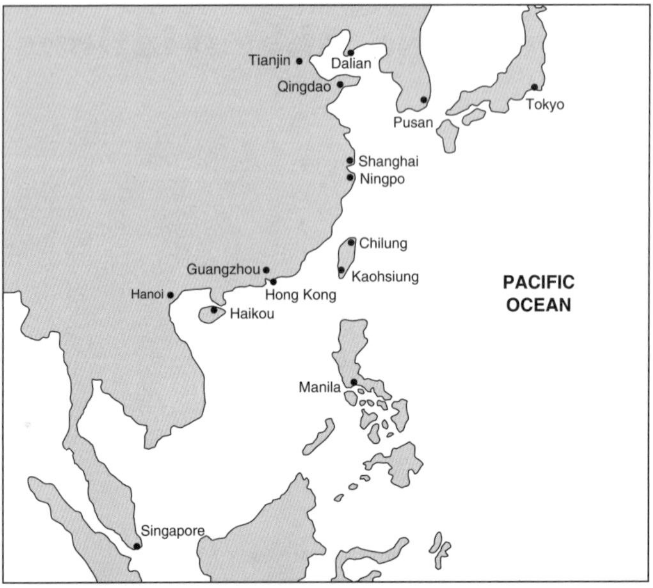

East and Southeast Asia Region

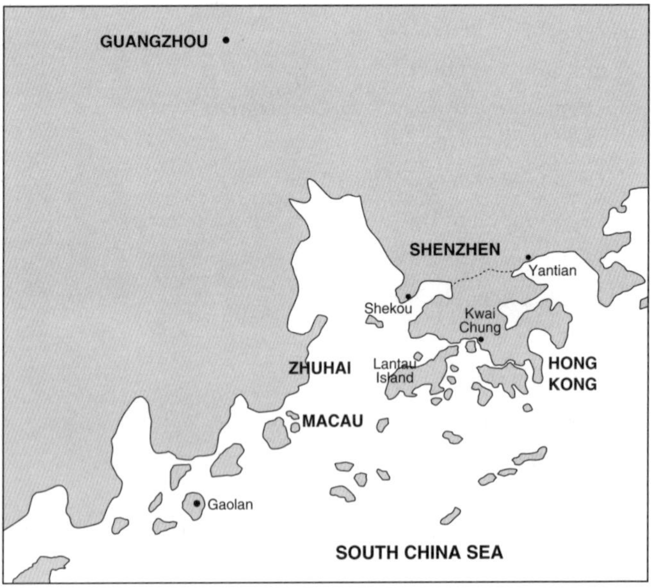

The Pearl River Delta Region

CHAPTER 1

Introduction

Hong Kong is a major trading economy and the most important entrepôt in the world. According to data released by the World Trade Organization (WTO) the export volume of Hong Kong during 1995 was US$173.8 billion, about 3.5% of the world's exports.[1] In that year, Hong Kong was ranked the ninth-largest exporter in the world. With an import volume of US$197.6 billion, or 3.8% of the world's total, it was ranked the seventh-largest importer then. If countries of the European Union were treated as a single economic entity, then Hong Kong's ranking would move to the fifth and the fourth positions for exports and imports, respectively. The role of the territory as an entrepôt is underscored by the fact that in 1995 its *domestic* exports accounted for a mere 17.2% of its total exports (i.e., domestic exports plus re-exports), and that only 27.2% of its imports were retained for domestic consumption and production.

A predominant share of Hong Kong's trade goes through its port. Moreover, some of the cargoes shipped through the port are not recorded as the territory's trade. Such cargoes are transshipments that pass through the territory without their ownership ever being transferred to any entity in Hong Kong. Thus, Hong Kong trade statistics do not fully reflect the importance of Hong Kong which operates the busiest container port in the world.

By providing efficient links to foreign markets, the port of Hong Kong has been instrumental in the territory's rapid economic development in the last several decades. At this point in time, there are three important sets of questions about the port of Hong Kong in the future.

1. Should Hong Kong continue to build additional container terminals given the development of other container ports in the region? If so, how many more should it build, and how soon?
2. What is the state of competition in the container handling industry, and how does it affect the costs of shipping, the welfare of local shippers, and Hong Kong's overall economic welfare?
3. Are there problems with the Port Development Board (PDB)'s "trigger point mechanism" as a way of determining the need for new terminals? How would the mechanism perform under increased competition? What factors other than market conditions should be considered?

In regard to the first set of questions, supporters for additional terminals argue that Hong Kong needs many more container terminals in order to meet the needs of additional facilities due to future expansion of cargoes. Some argue that rapid expansion of the container port is necessary to defend the territory's leading position and to pre-empt the development of potential competitors. Opponents of rapid development of container terminals argue that the future expansion of cargoes will be substantially slower than it has been in the last decade due to (a) the rapid container port development in South China and Taiwan, and (b) Hong Kong's high land and labour costs vis-a-vis regional competitors.

Our position is that while expansion would probably slow in the coming decade, the threat of potential competitors such as Yantian in Shenzhen may also take time to materialize. However, judging from recent statistics, this is happening sooner than is commonly expected. Thus, the slowdown in cargo expansion and the prospect of potential competition do not necessarily imply that there is no need to build new terminals (beyond Terminal No. 9).

In regard to the second set of questions, shippers and shipping lines argue that terminal operators are a virtual monopoly. Yet terminal operators counter-argue that they engage in fierce competition for customers. The PDB argues that there is more competition in Hong Kong's container handling industry than there is in most other ports in the world. Thus, while shippers and shipping lines would like to see the addition of new terminals and operators, the existing operators and the PDB do not see much need to change the status quo. Not unexpectedly, the interests of shippers, shipping lines and terminal operators are inherently in conflict.

Our position is that while there are currently four terminal operators in Hong Kong's container port, competition exists mainly between only two: Hongkong International Terminals (HIT) and Modern Terminals Limited (MTL). Given the small number of independent major operators, the competition among the existing four operators is less intense than it would be if there were additional independent competitors. Under the existing conditions, the fees charged by terminal operators' (terminal "tariffs") would be closer to the joint profit-maximizing (or cartel) tariffs than to competitive tariffs, which allow operators just to break even. Higher terminal tariffs would definitely hurt the interests of shipping lines and shippers. From the point of view of Hong Kong's overall economic welfare, however, competitive tariffs are not in the territory's best interests, either.

In regard to the third set of questions, supporters of the "trigger point mechanism"[2] see it as a compromise between the need to protect the investment of private terminal owners and operators on one hand, and the need to meet increasing demand on the other. However, critics have pointed out that high tariffs discourage demand and slow down the expansion of terminals.

Our position is that the functioning of the mechanism depends on whether the degree of competitiveness of the container handling industry is socially optimal. If the industry is optimally competitive, then the mechanism will be efficient in meeting new demand. But if it is not sufficiently competitive, then tariffs will be too high and demand will be under realized, thus leading to under expansion of terminals. Conversely, if the industry is excessively competitive, then tariffs will be too low and demand will be over realized, resulting in over expansion of terminals. We recognize that because many of the terminals in the world are subsidized by government to varying degrees, therefore, Hong Kong terminals compete from a disadvantageous position and their scale of operation is necessarily less than what it would have been if all terminals have to charge tariffs based on true costs.

In the following chapter, the historical development of Hong Kong's port and the role of the PDB in planning and developing port facilities will be briefly examined. Chapter 3 will describe the major features of the territory's container handling industry, and Chapter 4 will analyze in some detail the industry's structure, firm conduct, and economic performance. The development of regional ports and their implications for Hong Kong will be discussed in Chapter 5. The capacity of existing and planned facilities will be estimated in Chapter 6 while future demand for the territory's container handling services in the next decade will be assessed in Chapter 7. In Chapter 8 the main findings of the two preceding chapters will be brought to bear on a key policy question: whether and when Hong Kong should build new terminals. Finally, conclusions will be drawn and recommendations made in Chapter 9.

Notes

1. WTO Press Release, 22 March 1996.
2. The "trigger point mechanism" is central to the planning of the number of berths for Hong Kong's container terminals. According to the Port Development Board, "new berths were not triggered until forecast throughput equalled working capacity of existing and planned berths. This caused berth supply to lag one berth behind demand." (PDB 1992, p. 23) On how this works, see the last section of Chapter 2; on the problems with the mechanism, see the last section of Chapter 8.

CHAPTER 2

The Port of Hong Kong: Recent Developments

In this chapter we first provide a very brief historical background of the port of Hong Kong. Then we examine the importance of port cargoes in the overall movement of goods by all modes of transport into and out of the territory. We also examine the breakdow of port cargo types. We follow that with a presentation of a breakdown of ocean and river cargoes according to the geographical location of the territory's trading partners. We focus on those cargoes travelling between Hong Kong and China. Given that Hong Kong is largely a container port, we analyze the evolution of its container throughput. Finally, we briefly describe the role of the Port Development Board (PDB) in the territory's port development.

Brief Historical Background [1]

After the Opium War of 1840, Hong Kong was ceded to Britain by the Treaty of Nanjing signed on 29 August 1842. Henry Pottinger, the first governor of Hong Kong, declared in March of that year that it would be a free port, open without discrimination to all ships. By the turn of the nineteenth century it became a major port for world trade. For the last one hundred years it has prospered as a free port.

The development of Hong Kong's container port came later. To meet the challenges of the new container era and to reap the opportunities it presented, a Container Committee was set up in

July 1966. On 2 December of that year, the committee presented its report with a recommendation to build container terminals in Kwai Chung. The first container ship arrived in the new terminal port in September 1972. At present, there are a total of eight terminals with eighteen ship berths and two barge berths in Hong Kong. Hong Kong became the number one container port in the world for the first time in 1987 and has ever since held on to that position, except in 1990 and 1991, when it lost its place to Singapore.[2]

Port Cargoes*

The volumes of freight moved through the port from 1983 to 1995 are given in Table 2.1 (see Appendix, p. 124). The total volume of ocean and river freight increased from 37 million tonnes in 1983 to 156 million tonnes in 1995. With the exception of 1989 and 1990, the annual growth rate during those years was in the double digits. Goods that passed through the port accounted for close to 90% of all freight movement, i.e., freight moved by ocean, river, rail, road and air.

In every year during the period the volume of goods arriving in the territory exceeded that of goods leaving it. That is to be expected, because a portion of the goods was retained by the resource-poor territory for its own consumption and production.

Ocean Cargoes and River Cargoes

Port cargoes have traditionally been divided into "ocean cargoes" and "river cargoes". In 1992 a new definition of these two types of cargoes was adopted. Under this definition, river cargo refers to cargo "carried by vessels plying between Hong Kong and all ports in the Pearl River (including Macau)."[3] Partly due to the change in definition and to a more systematic effort to collect river cargo data, the volume of such cargoes increased sharply in that year. The slowdown in the growth of ocean cargoes from 1991 to 1992 can

---

* All tables in Chapter 2 are given in the Appendix starting on page 124.

Figure 2.1
Freight Movement through the Port of Hong Kong, 1983–1995

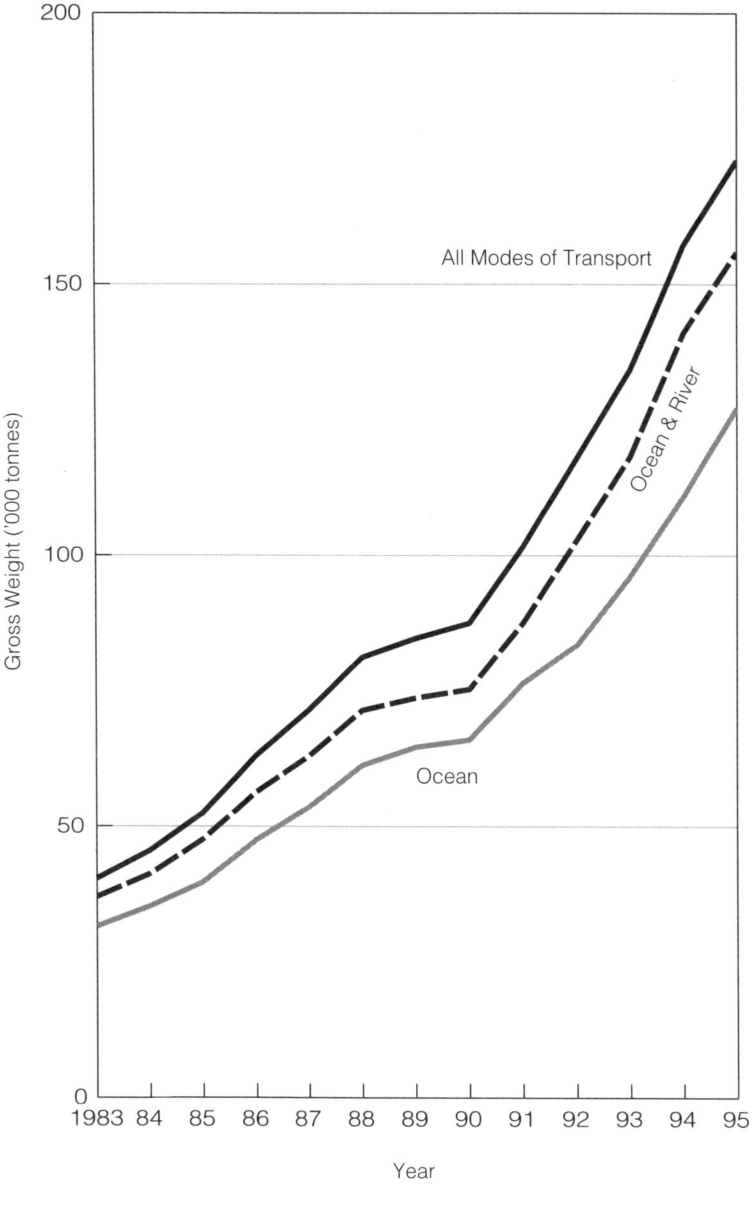

perhaps be attributed in part to the adoption of this new definition. The time paths of ocean cargo, ocean plus river cargo, and total freight movement into and out of the territory are depicted in Figure 2.1.

The share of ocean cargoes in port cargoes (ocean plus river cargoes) was slightly larger in inward cargoes (83% to 89%) than it was in outward cargoes (71% to 86%). Judging by their annual growth rates, it is clear that over the period the volume of river cargoes experienced greater fluctuations than did that of ocean cargoes.

Direct Shipment and Transshipment

Cargoes are also distinguished by whether or not they are part of the territory's foreign trade. "Direct shipment" is cargo imported into the territory or exported from it, whereas "transshipment" refers to cargoes shipped from one place to another place through Hong Kong without at any time conferring ownership of the cargoes upon any entity in the territory. In other words, transshipment cargoes are not recorded as its imports, exports or re-exports.

In inward ocean cargoes, the share of transshipment accounted for 9.4% of such cargoes in 1983, and rose steadily to 18.1% in 1995,[4] implying that the share of direct shipment dropped steadily to 81.9% in 1995. The share of transshipment in outward ocean cargo was much higher: 38.2% in 1983 and 42.6% in 1995. Combining inward and outward cargoes, the share of transshipment rose from 16.2% in 1983 to 25.8% in 1995.

In inward river cargoes, the share of transshipment rose from 7.2% in 1992 to 16.6% in 1994. The corresponding figures for outward river cargoes were 16% and 31.5%, respectively, and those for total river cargoes were 10.7% and 23.5%, respectively.

Thus, statistics for both ocean and river cargoes imply that transshipment has become increasingly more important relative to direct shipment. Transshipment is a testimony to the growing importance of Hong Kong's role as an entrepôt. In 1995 transshipment for ocean cargoes accounted for slightly more than one quarter of Hong Kong's total cargoes, while for river cargoes it accounted for slightly less than one quarter.

Break Bulk, Dry Bulk, Liquid Bulk, and Containerized Cargoes

Port cargoes can be divided into four different cargo types according to how they are physically packaged and moved and by which type of ship. These four types are break bulk, dry bulk, liquid bulk, and containerized cargoes.[5] Hong Kong's largest dry bulk cargoes are coal imports by Hongkong Electric and China Light and Power. The two companies operate their own berths for the discharging of coal. Cement is another important kind of dry bulk cargo. The three major cement companies (China Cement, Green Island Cement, and Far East Cement) operate their own terminals in Tap Shek Kok, Tsing Yi and Lamma Island, respectively.

The most important liquid bulk cargoes are petroleum products which are delivered to the oil suppliers (China Resources, Caltex, Esso, Feoso, Mobil, and Shell) at their specialized terminals and piers in Tsing Yi.

Break bulk is handled mid-stream or at general purpose piers, while containerized cargoes are handled mostly at dedicated container berths or in mid-stream.

The volume of each type of direct ocean cargo from 1987 to 1994 according to inward cargoes, outward cargoes, and their sum is given in Table 2.2. As the figures show, in inward cargoes the combined share of dry bulk and break bulk in direct ocean traffic fell from 56% in 1987 to 34% in 1994. While liquid bulk's share increased moderately, from 18% in 1987 to 22% in 1994, that of containerized cargoes jumped from 26% to 43% over the same period. In the case of outward cargoes, the shares of these four cargo types underwent similar changes, but the share of containerized cargoes was much higher (72% in 1987 and 80% in 1994), and the shares of the other three types were much lower throughout the period.

As a result of these changes in the shares of cargo types, the shares of break bulk, dry bulk, liquid bulk, and containerized cargo in 1994s direct ocean cargoes were 8%, 21%, 19%, and 52%, respectively.

Table 2.3 is similar to Table 2.2 except that the former deals with transshipment ocean cargoes. The growth in containerized cargoes at the expense of the other cargo types was even more dramatic than was that for direct ocean traffic. Containerized cargo's share in transshipment ocean traffic rose from 71% in 1987 to an incredible 95% in 1994.

Table 2.4 provides similar information for river traffic from 1992 to 1995. While break bulk and dry bulk accounted for over 64% of direct river traffic in 1995, 81.7% of transshipment river traffic was accounted for by containerized cargoes. Even though the relative importance of transshipment was catching up, its volume was only slightly more than one quarter of that of direct shipment in 1995. As a result, in that year containerized cargo accounted for only 31.7% of all river traffic.

The shares of cargo types in ocean traffic in 1994 and in river trade in 1995 are depicted in Figure 2.2. It is clear that if liquid and dry bulk cargoes (which are handled at specialized terminals and piers) are left aside, then the lion's share of all other cargoes is containerized. In other words, Hong Kong has become mainly a container port. Moreover, the degree of containerization, defined as containerized cargoes divided by all cargoes, of its cargoes has been increasing and is expected to increase further. Consequently, in our study we shall pay particular attention to the container handling industry.

Cargoes by Geographical Zone

We have grouped Hong Kong's trading partners into ten geographical zones including a residual category of "others". The volume and percentage share of inward direct ocean cargoes by zone are given in Table 2.5. In 1995 the largest share of inward direct ocean cargoes came from ASEAN, to be followed by Japan and Taiwan, while China came in only seventh. Table 2.6 contains

Figure 2.2
Shares of Cargo Types

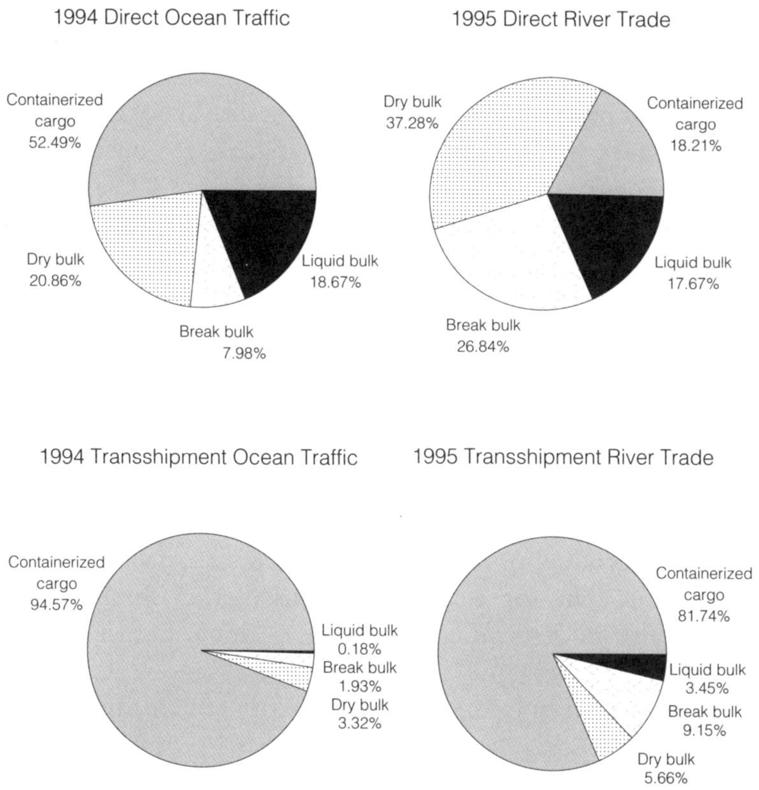

similar information for outward direct ocean cargoes. The top four regions in descending order in 1995 were North America, China, Europe, and ASEAN.

The volume and percentage shares of inward and outward containerized direct ocean cargoes are given in Table 2.7 and Table 2.8, respectively. The degree of containerization can be obtained by dividing the volume of containerized cargoes in these two tables by that of direct ocean cargoes given in Table 2.5 and Table 2.6. With the exception of China and Taiwan in some years, each region's degree of containerization of outward cargoes exceeded that of its inward cargoes every year. These statistics reflect the fact that: (a) Hong Kong was a producer of manufactured goods, (b) that it imported raw materials and semi-finished products while exporting finished products, and (c) that it was a provider of services that did not result in cargoes.

In 1995 the average degree of containerization of inward direct ocean cargoes was 43.7%, but it was 78.2% for outward direct ocean cargoes.[6] In the same year, only two regions' (Taiwan's and North America's) degree of containerization of inward cargoes exceeded 85%. In contrast, for outward cargoes, the degree of containerization for all regions except China and Taiwan exceeded 88%, while that for Australasia/Oceania, North America, and Europe exceeded 99%.

The volume and percentage shares of inward transshipment ocean cargoes by zone are given in Table 2.9; those of outward transshipment ocean cargoes are given in Table 2.10. In 1995 the top four zones that supplied transshipment ocean cargoes to the port, in descending order, were China, North America, ASEAN, and Taiwan. The top four destinations of transshipment ocean cargoes, also in descending order, were China, Europe, ASEAN, and North America.

The volume and percentage shares of inward and outward transshipment ocean containerized cargoes by zone are given in Table 2.11 and Table 2.12, respectively. Dividing these figures by those provided in Table 2.9 and Table 2.10, we can obtain the degree of containerization of transshipment cargoes. While in most cases the degree of containerization of outward cargoes exceeded that of inward cargoes, both were rising, and there was no significant difference between the average degree across all regions. In 1994 the average degree of containerization of total outward transshipment ocean cargoes was 97.7%, as compared to 91.6% for inward cargoes, and in 1995 the corresponding figures were 99.1% and 96.5%. In any event, these figures clearly show that transshipment ocean cargoes were predominantly containerized.

Given the new definition adopted for river trade in 1992, it is to be expected that the lion's share of Hong Kong's river trade has been trade with China. Detailed statistics about the origin and destination of river trade will be provided in the next section.

Cargoes between China and Hong Kong

Relative Importance of the China–Hong Kong Port Cargoes

A major component of Hong Kong's port cargoes is freight movement between China and the territory. The relevant information is provided in Table 2.13.

In the freight movement between Hong Kong and China, port cargoes accounted for 61% to 75% of freight movement by all modes of transport throughout the period 1983–1995. These shares were substantially lower than were the shares of the territory's total port cargoes in its total freight movements by all modes of transport. In addition, the ratio of ocean to port cargoes between the territory and China (37% to 56%) throughout the same period was also substantially lower than was the average ratio for all the territory's trading partners.

These two differences are to be expected because Hong Kong is located on the coast of China (so they are connected by land as well as by sea and air), while Hong Kong's non-China trading partners and the territory are separated by seas and oceans. Not surprisingly, non-China ocean cargoes accounted for over 96% of all non-China port cargoes in every year from 1983 to 1994.[7]

Given the distance-based definition of river-trade cargoes beginning in 1992, it also comes as no surprise to know that Guangdong had a predominant share during the period, whereas Macau and Guangxi had only a relatively small share. In 1994 the latter two areas together accounted for 4.6% of Hong Kong's inward river cargoes, 24.6% of its outward river cargoes, and 13.9% of its total river cargoes.[8] The 1995 figures fell to 4.7%, 13.7%, and 9.1%, respectively.

How important were the port cargoes in goods movement between Hong Kong and China? According to Table 2.14, from

1983 to 1995 the share of port cargoes from China in the territory's total inward port cargo fluctuated between 23.4% and 27.4%. The 24.55% share in 1995 was not much different from that of a decade earlier. The situation of outward cargoes was significantly different. During the same period, the share of port cargo to China in the territory's total outward port cargo rose from 12.5% in 1983 to 40% in 1995. Despite some reversals, the overall trend was clearly up. Combining inward and outward cargoes, China port cargoes accounted for about one quarter of the territory's total port cargoes from 1983 to 1991; beginning in 1992 the figure rose to about 30%.

Geographical Breakdown

Table 2.15 provides a breakdown of China-related ocean traffic according to the geographical location of the ports of loading and discharge. In every year from 1985 to 1994, North China ports provided more inward cargoes than did Middle China ports, which in turn provided more inward cargoes than did South China ports.

In the case of outward cargoes, however, there was a shift in the regions' relative importance. From 1985 to 1988 the ranking in descending order was Middle China, North China, and South China. From 1989 to 1991 South China took second place from North China, and beginning in 1992 South China became the leader. Basically, the same underlying force was at work: Hong Kong's trade with South China continued to expand at a faster pace than did trade with the other two Chinese regions.

Combining inward and outward traffic, South China's share increased from 18% in 1985 to 31% in 1994. South China's growing importance came at the expense of both North and Middle China.[9]

Table 2.16 shows the volume of trade of six Chinese regions, regardless of whether the goods were shipped through the territory. Figures in the table were compiled using statistics provided in *China Customs Yearbooks* from 1992 to 1994.[10] The Northeast and the Northwest lost their shares in exports, while the South and — to a lesser extent — the East gained. The other two regions maintained similar shares over the period. In terms of imports, the shares of the North, the Southwest, and the Northwest gained at the expense of the South and the East. In total trade, the South and Southwest gained slightly at the expense of the Northeast.

The containerized part of Hong Kong–China ocean cargoes behaved differently. As shown in Table 2.17, in 1994 Middle China accounted for the largest share of containerized cargoes, followed by South China and then North China. Middle China was most important in inward cargoes to the territory, but its leading position in outward cargoes from the territory was taken over by South China beginning in 1993. In 1994 ports in Middle China accounted for 40% of the ocean traffic, those in South China accounted for 29%, and those in North China accounted for 17%.

The river-trade cargoes from 1992 to 1995 by ports and by shipment types are given in Table 2.18. All the twelve individual ports listed in the table were in the Pearl River Delta, to be followed by Guangxi Province, Macau, and ports whose identity was not made known by the government agencies involved in data collection. As one would expect, river-trade cargoes to and from China were predominantly cargoes to and from the Pearl River Delta. Guangxi's river-trade cargo accounted for only 4.8% of the total in 1992 and 2.6% in 1995.[11] Relatively speaking, Guangxi's share in direct river cargo exceeded its share in transshipment river cargo every year since 1992.

Within the Pearl River Delta, the top three ports with the largest shares of direct river cargoes were, in descending order, Zhuhai, Shenzhen, and Guangzhou. In transshipment, Guangzhou was by far the most important, while Zhuhai and Shenzhen competed for the second place. In terms of total shipment, the order was Zhuhai, Guangzhou, and Shenzhen, except in 1993 when Guangzhou and Shenzhen switched positions.

Upon comparing Table 2.18 with the last three columns of Table 2.15, it is clear that ocean traffic of the various Chinese regions paled against total river traffic, which, by definition, was

Figure 2.3
Hong Kong Overall Container Throughput, 1985–1995

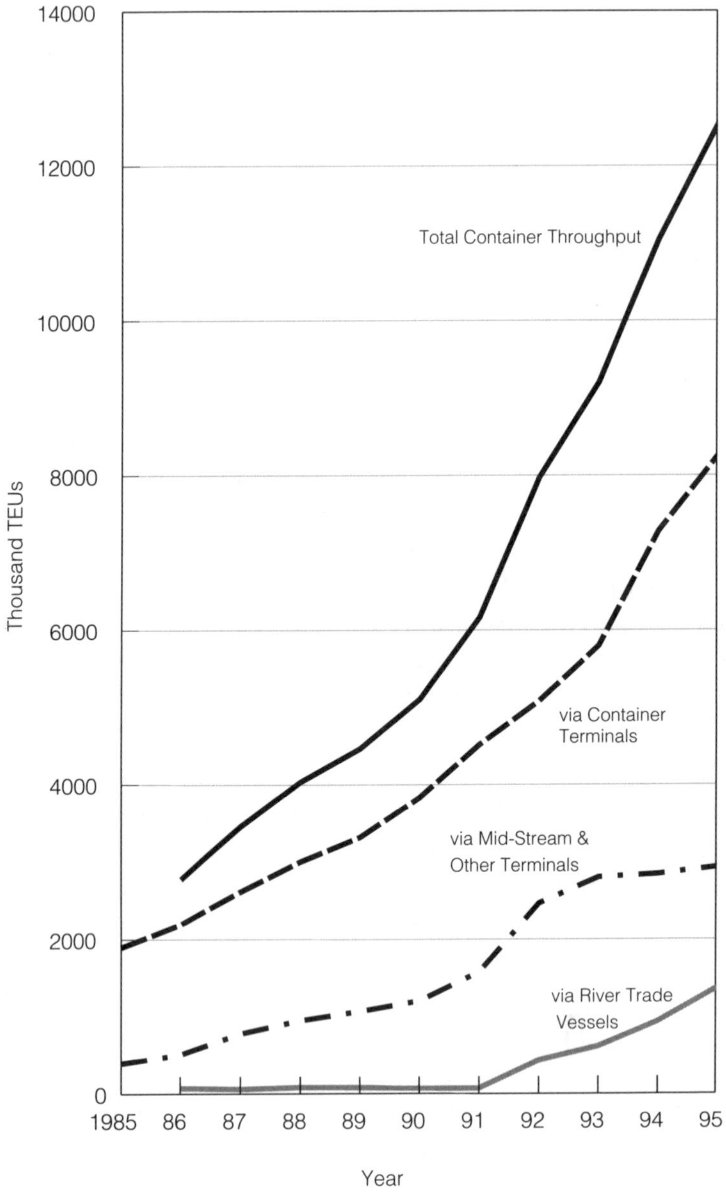

under South China. Thus, South China's port traffic dominated that of all other Chinese regions.

As indicated in Table 2.4, the degree of containerization of transshipment river-trade cargoes was four to six times that of direct shipment river-trade cargoes, and the faster growth of the former resulted in an increase in the overall degree of containerization, from 17.6% in 1992 to 31.7% in 1995.

Container Throughput

Hong Kong's container throughput volumes as handled by the container terminals in Kwai Chung, mid-stream operators, and river-trade vessels are provided in Table 2.19. The unit of measurement is Twenty Foot Equivalent Units (TEUs). The throughput figures for the three kinds of operators and their sum total from 1985 to 1995 are depicted in Figure 2.3.

Table 2.19 shows that inward river traffic was more or less equal to outward river traffic throughout the years. While container terminals experienced imbalances in some years, such imbalances were more or less offset by opposite imbalances of mid-stream operation. As a result, total inward container traffic roughly equaled to total outward container traffic throughout the period.

Over time the container terminals have experienced a decline in their shares, from 79% in 1986 to 66% in 1995. The decline was most drastic between 1991 and 1992. Adopting a new definition for river trade and using a new system to collect river-trade data could be part of the underlying reasons for this decline. Indeed, the TEUs handled by river vessels jumped by six times, and their market share jumped by four times between the two years. However, the change of definition alone could not have explained the whole story, because the share of mid-stream operation went up by more than five percentage points during the same period. One other explanation is the development of delta ports and the acceptance of river barging as an alternative to trucking as a means of moving cargo. Another explanation is the lack of handling capacity in the years before the new Terminal 8 was completed and before investment to enhance the existing terminals' design capacity bore fruit.

Throughout the entire period, mid-stream operations enjoyed a significant share of the market. With a volume of about half a million TEUs in 1986, the throughput handled in mid-stream reached almost three million TEUs in 1995. Its share has been above 20% since 1987; it rose to 30.9% in 1992 before dropping back to 23.3% in 1995.

The throughput of river vessels was 432,000 TEUs in 1992 when a new definition for river trade was first adopted, and when a new system was first used to collect data more systematically. The throughput quickly rose to 1.36 million TEUs in 1995. As a result of its relatively faster growth, the market share of river trade increased from 5.4% in 1992 to 10.9% in 1995.

Table 2.20 gives the percentage shares of transshipment containers in the total container throughput of the three types of operators. For the terminal operators in Kwai Chung, the percentage continued to decline, from 27.2% in 1985 to 14.6% in 1993, but the trend reversed in the last two years of the period. In 1995 transshipment containers accounted for 27.3% of total throughput, almost exactly the same percentage as a decade earlier. The dramatic increase in throughput probably led the PDB to accelerate the requirement for additional container terminals.

Similar data for mid-stream and river trade became available only beginning in 1992. The percentage of transshipment for mid-stream was definitely smaller, varying from 6.6% to 9.7%, but that for river-trade vessels varied narrowly, from 17.6% 19.3%.

Ignoring the identity of the container handlers, 22.2% of all containers handled in Hong Kong in 1995 were transshipment containers. It should be noted that some of the containers handled were empty. The percentage of empty containers ranged from a low of 14.5% in 1994 to a high of 18.1% in 1986, he first year such information became available for all three types of operators.[12] In 1995 empty containers made up 16.2% of the total container throughput. Throughout the entire period, river vessels accounted for the highest percentage, and after 1989 mid-stream accounted for the lowest.

Port Development Board (PDB)

As explained in its Annual Reports, the PDB was formed in April 1990 to advise the Governor, through the Secretary for Economic Services, on all aspects of planning and development for the port of Hong Kong. A crucial role of the PDB is "to recommend strategies for creating new port facilities and to co-ordinate government and private sector involvement in developing them." Hong Kong is essentially a container port because (a) break bulk cargoes have virtually disappeared and (b) dry bulk and liquid bulk cargoes are handled at special purpose terminals with specialized facilities. The PDB does not get involved with such specialized facilities.

Given the premise of relying on the private sector to provide cargo-handling facilities, the PDB could be regarded as an institutional arrangement wherein the interests of shippers, shipping lines and container handlers are reconciled and balanced. Central to the planning of container terminals is the *trigger point mechanism,* under which "new berths were not triggered until forecast throughput equalled working capacity of existing and planned berths. This caused berth supply to lag one berth behind demand." (PDB 1992, p. 23)

Recently the container handling Committee under the PDB decided to "(a) continue the existing five year period for projections, (b) introduce a seven-year projection, and (c) introduce a ten-year 'priming' projection." (PDB 1995, p. 16)[13]

To provide additional sites for mid-stream operations, the government has already set aside land on Stonecutters Island, where a permanent site for use in 1998 is to be built.[14] Other potential sites for mid-stream operations include Lung Kwu Tan, to the west of Tuen Mun; Tseung Kwan O; Northshore Lantau; and east of Terminals 10 and 12 in Northeast Lantau.

To accommodate the rapid expansion of river trade, a dedicated terminal will be built in Tuen Mun, and part of the facility is expected to become available for use in 1999. A possible site for a second such terminal has been identified at Northshore Lantau. In addition, some barge berths will be built right next to container ship berths in order to reduce double handling of containers (brought to river-trade terminals and then sent to the main container terminals located elsewhere). As a result of the swap of ship berths in Kwai Chung and Tsing Yi to resolve the political controversy surrounding the award of Terminal 9, two additional berths to allow greater use of barges have been added to the new terminal.

With the above brief historical background of the port of Hong Kong, we are now ready to turn to the territory's container handling industry. In the next two chapters, we shall analyze the industry in some depth. Our analysis will cover industry structure, firm conduct, and economic performance.

Notes

1. The information in this section is based on Taylor (1991, Chap. 9).
2. See Rodwell (1989, p. 56) and David K.Y. Chu, "Challenges to the Port of Hong Kong Before and After 1997," *Chinese Environment and Development*, Vol. 5 (1994), No. 3, pp. 5–23.
3. Prior to January 1993, ocean-going vessels were defined as vessels completing port formalities at the Marine Department's Port Formalities Office, while vessels completing port formalities at the department's District Marine Offices were defined as river vessels. Under the new statistical system, river trade vessels are defined as vessels that run between Hong Kong and the Pearl River as well as other inland waterways in Guangdong and Guangxi (see a discussion under Section 4.2.4 of Port Development Board (1996)). As the footnotes to the various tables have indicated, however, the ocean and river cargo figures in 1992 were reported under the new statistical system.
4. The 1983 figures can be found in Table C1 of the PDB's *Port Cargo Forecasts 1993*, while the 1995 figures were obtained from *Hong Kong Shipping Statistics*, October to December 1995, Tables C1 and C2.
5. In terms of ship types classified by the Census and Statistics Department, break bulk includes cargoes carried by conventional cargo vessels and "other" ships. Dry bulk includes cargoes carried by dry bulk carriers and

log (timber) carriers. Liquid bulk includes cargoes carried by oil tankers, gas carriers (tankers), chemical carriers (tankers), and Ore-Bulk-Oil (OBO) carriers. Finally, containerized cargoes include cargoes carried by container and semi-container vessels.

6. The figures were obtained from *Hong Kong Shipping Statistics*, October to December 1995, Tables C1, C2, C3, and C4.

7. See Tables Bl and B2 of *Port Cargo Forecasts 1993* and Table B3 of *Port Cargo Forecasts 1995* for data on ocean and river cargoes.

8. Guangxi and Macaus' combined share in Hong Kong's total inward river trade cargoes in 1992 and 1993 was 8% and 6.6%, respectively. Their combined share in outward river trade cargoes in the two years was 15.5% and 14%, respectively.

9. The combined share of North and Middle China fell from 73.3% in 1985 to 50.1% in 1994.

10. The trade figures provided by the Customs Administration differed somewhat from those provided by the MOFTEC. We use the former because we are interested in where goods were shipped in and out of China, and presumably statistics provided by customs would be more reliable in that particular regard.

11. The shares in 1993 and 1994 were 2.5% and 7.1%, respectively.

12. For data up to 1994, see Table E7 in Port Development Board (February 1996); 1995 figures were obtained from the Marine Department.

13. The same report went on to state the following. "Under the mechanism, planning was triggered on CTs 12 and 13. Present forecasts predict the need for the first berths of these terminals to come on line in August 2000 and August 2003, respectively." According to the PDB, the government has completed the condition of grant for CT 10, and the relevant seabeds are ready for bidding in the near future. The condition of grant for CT 11 will be the focus of future work.

14. See Planning Department and Port Development Board's (1995) "Outline Implementation Programme" for information about the planning of future port facilities.

CHAPTER 3

Hong Kong's Container Handling Industry

We have already seen that Hong Kong is primarily a container port.[1] Its container handling industry can be divided into (a) container terminal operators who now handle almost two-thirds of containers, (b) mid-stream operators who now handle not quite one quarter of containers, and (c) river-trade operators who control about 11% of the market. In the following sections, we shall analyze them in the above order. After the three principal segments of the container handling industry are described, we shall then examine their substitutability and complementarity.

Container Terminals

All the existing container terminals in Hong Kong (Terminals 1 to 8) are located in Kwai Chung. Before the terminals in Kwai Chung were created, container handling was performed at Kowloon Wharves, North Point Wharf, and Kowloon Docks, primarily using the container ships' own cargo-handling equipment.[2]

In the first phase of Kwai Chung's development as a container port, the development right of Terminals 1 to 3 was awarded on 18 August 1970 to three companies owned by shipping lines.[3] Terminal 1 went to Modern Terminals Limited (MTL), which was established in 1969 by a consortium led by three shipping lines: Overseas Container Lines (OCL), Ben Line, and Hapag-Lloyd.[4] Terminal 2 went to Kowloon Container Warehouse, a subsidiary of Japan's Oyama Shipping. Terminal 3 went to Sea-Land Orient, a subsidiary of the U.S. shipping line Sea-Land. The construction of these three terminals began in 1970 and was completed in 1973. On 5 September 1972 OCL's Tokyo Bay became the first-ever container ship to visit Kwai Chung.

Additional terminals were subsequently built to meet demand. The most recently built terminal was Terminal 8. The east portion of Terminal 8 became fully operational in July 1994, and the west portion became fully operational in March 1995.

At present, there are a total of four terminal operators in Kwai Chung. Hongkong International Terminals (HIT) is the world's largest privately owned container terminal operator; it owns Terminals 4, 6 and 7. The second-largest operator in Kwai Chung is MTL which owns Terminals 1, 2, 5 and Terminal 8-West. The third and newest operator is COSCO-HIT (see next page) which owns two berths on Terminal 8-East. The fourth and smallest operator is Sea-Land Orient Terminals (SLOT) which owns a single berth on Terminal 3.

HIT was set up in 1974 by the Hongkong and Whampoa Dock Company (HWD), which in 1977 became part of Hutchison Whampoa. In 1969 the Dock Company was asked by Orient Overseas Container Line (OOCL) to provide berthing facilities for its container vessels. In response, the Dock Company set up "a small specialist yard at Hunghom based on a 180-metre berth and an adjoining container storage area of some 9,000 square metres". container handling operations started in October 1969 and later expanded to North Point and Kwun Tong.[5]

HIT started the construction of Terminal 4 in 1974, and the terminal was fully operational by 1976.[6] In February 1976 HIT purchased Terminal 2 from Oyama Shipping.[7] During the second phase of Kwai Chung's port development, HIT went on to build Terminal 6 despite a general lack of interest by investors. As part of an agreement to secure the right to develop all three berths on Terminal 6, it agreed to sell Terminal 2 to MTL in 1985. The sale took place in 1988 and Terminal 6 became operational later in the same year. In that year HIT also acquired the development rights to Terminal 7. The operator's historically high bid of $4.39 billion sent shock waves throughout the industry.

Since the numbering of terminals is not exactly related to location, Terminals 4, 6 and 7 are in fact contiguous. With a total berth length of 3,292 metres and a total area of 89 hectares, HIT operates a total of ten ship berths and one barge berth with a stacking capacity of 57,000 TEUs. In addition Hutchison also has a Distribution Centre with over 293,000 square metres of space.

MTL completed Terminal 1 in 1972 and then acquired the rights to build Terminal 5 in 1976 during the second phase of Kwai Chung's development. In 1988 it bought Terminal 2 from HIT, a strategic horse trading with HIT that left MTL with three contiguous terminals. In May 1992 China Merchants acquired a 15% stake in MTL.[8] Most recently MTL's Terminal 8-West became fully operational in March 1995.

With a total of 1,822 metres of berth length and a total area of 85 hectares, MTL operates five ship berths with a stacking capacity of close to 50,000 TEUs.[9] In addition, it has close to 110,000 square metres of storage space and about 12,000 square metres of Container Freight Station facilities.

As a reflection of the growing importance of China business to the port of Hong Kong, in 1991 Terminal 8-East was awarded to a joint venture formed by HIT and COSCO-Pacific, namely, COSCO-HIT. It became fully operational in July 1994. With 640 metres of berth length, 380 metres of barge berthing and an area of 30 hectares, it operates two ship berths and one barge berth and has a stacking capacity of close to 20,000 TEUs.

SLOT was set up in 1991, succeeding Sea-Land Orient Ltd. In the same year, it reached an agreement with HIT to use the adjacent Terminal 4 with priority. Its only berth is 305 metres long. With an area of 16.7 hectares, its container yard has a stacking capacity of close to 7,000 TEUs. It operates the world's largest container freight centre, Asia Terminals Centre, which has an area of 800,000 square metres of space, of which close to 15,000 square metres are used for a container freight station and over 640,000 square metres for storage.

In 1995 the throughput of HIT was 3.96 million TEUs,

representing 48% of Kwai Chung's total throughput. In the same year, MTL handled 2.11 million TEUs with a share of 25.6%. COSCO-HIT handled 1.19 million TEUs with a share of 14.4%, while SLOT handled 0.99 million TEUs with a share of 12%.

The next new terminal, Terminal 9, will be located on Tsing Yi Island across from Kwai Chung. There was a political controversy surrounding the awarding of two berths on Terminal 9 to the Tsing Yi Consortium led by Jardine Matheson. The controversy was resolved after the existing and future terminal operators were instructed in January 1996 to swap their berths. The swapping carried with an implicit understanding that the Tsing Yi Consortium would move out of Terminal 9 in order to reduce the degree of scattered ownership of berths.

After months of negotiations, an agreement was finally reached in September 1996. Under this agreement, which was later endorsed by China via the Sino-British Joint Liaison Group, the Asia Container Terminals (ACT), a new consortium led by the Sea-Land Group,[10] would purchase the two berths on Terminal 8-West from MTL. In addition, SLOT would manage ACT. In return MTL would own three ship berths (i.e., one that was originally awarded to it plus two from the swap with ACT) and one barge berth on Terminal 9-South. HIT would take one ship berth from its original award plus one barge berth on Terminal 9-North.[11] Thus, relative to the original award of Terminal 9, both HIT and MTL received a bonus from the swap agreement in the form of a barge berth, which is similar to a regular ship berth except that it has much less backup land.

After the construction of Terminal 9 is completed and the swap agreement implemented, there will be five operators in the container port of Kwai Chung and Tsing Yi. The ownership structure will be as follows:

1. HIT owns a total of eleven ship berths and two barge berths on Terminals 4, 6, 7 and 9;
2. MTL owns a total of six ship berths and one barge berth on Terminals 1, 2, 5 and 9;

3. COSCO-HIT owns two ship berths and one barge berth on Terminal 8-East,
4. ACT owns two ship berths on Terminal 8-West, and
5. SLOT owns one ship berth on Terminal 3.

It should be noted that the above five terminal operators are not truly independent. The link of COSCO-HIT with HIT is obvious. Equally obvious is the link between SLOT and ACT.

Given the lack of waterfront land for further expansion of the container port berths in the Kwai Chung and Tsing Yi Island area, the Port Development Board (PDB) plans to build a new container port in Northeast Lantau. In the government's Port Development Strategy, four terminals (Terminals 10 to 13) with seventeen or more berths are envisioned.

Mid-Stream Operations

As described in a PDB leaflet, "mid-stream operations involve the loading and unloading of cargo ships mooring at buoys or anchorages in the harbour. A wide range of vessels from bulk, break-bulk, semi-container to fully cellular container ships are served."[12] Buoys provided by the Marine Department are safer than the ships' own anchors. Typhoon buoys are tied to the sea bottom and can ensure that ships are secured during a typhoon.

According to the PDB, by the early nineties there were "some 2,000 privately-owned lighters service ships moored mid-stream and about 200 lighters" specially designed to carry containers.[13] "About 72% of the cargoes handled by the mid-stream sector in 1994 were containerized. These amounted to 2.8 million TEU units or 25% of all containers handled in Hong Kong."[14] In 1995 more than 2.9 million TEUs, or 23% of all containers handled in Hong Kong, were handled in mid-stream. In these two years the container throughput, handled mid-stream, exceeded the container throughput of many container ports in the world.

Mid-stream operations rapidly developed in the mid-eighties for two reasons. First, the tariffs charged by the terminals were regarded as high by shipping lines. Second, trade with Southeast Asia expanded at a fast pace, giving a comparative advantage to mid-stream operations. As is indicated in Table 2.19, no river-trade container throughput data were available before 1985; we can obtain only the relative shares of terminals and mid-stream operations for the earlier period. In 1983 mid-stream operators accounted for 10.9% of the combined total of mid-stream and terminal throughput. Its share climbed steadily to reach a maximum in 1992 and 1993. In these two years, before Terminal 8 became available, the share of mid-stream operations in overall container throughput slightly exceeded 30%, or over 32% if river-trade throughput was excluded.

The advantage of mid-stream operations relative to container terminals' operations lies in the former's ability to operate simultaneously at both sides of a ship, as opposed to only one side at a time in the case of berths. Its disadvantage lies in the limitation of its equipment (cranes) in terms of how far they can reach and how fast they operate. Its cost advantage is therefore with smaller ships, because the berthing time of ships is more or less equal regardless of size, thus making the average cost per TEU high when smaller ships are serviced at the berths. The handling fees in mid-stream are 40% to 60% cheaper than they are at the terminals.[15] In contrast, mid-stream operations become technically impossible with very large ships.

In the early days of container handling in Hong Kong there were many small firms in mid-stream operations. Over time, however, major players moved into the business, because the operations became increasingly more capital-intensive. The tugs and lighters for mid-stream operations require heavy capital investment. Due to the temporary nature of the tenancy on the waterfront, however, small operators have no incentive to spend money to develop them.[16]

Terminal operators were unhappy with the presence of mid-stream operators who were seen to be competing against the terminal operators with subsidized facilities. This sentiment is summarized by Rodwell (1989, 75) as follows:

"There are differences of opinion between the Administration and the terminal operators (including HIT) over what is seen as the official subsidy of Hong Kong's other cargo handling areas. The terminal operators fail to understand the logic behind requiring the private development of Hong Kong's major port infrastructure while encouraging the persistence of what they regard as antediluvian methods in the cargo handling areas, methods which help to undercut the container tariffs by a wide margin."

A debate took place in the PDB about whether mid-stream operators should be allowed to compete against terminal operators, given that the latter made heavy capital investment whereas the former used public facilities at low cost. In the end it was decided that mid-stream operations served a useful function. They were regarded not only as a safety valve for overflow from the terminals when the latter reached their capacity limits but also as a competitive fringe putting pressure on terminal operators' pricing policies.

Realizing that mid-stream operations were here to stay, some terminal operators entered the game, as well. HIT has been in the mid-stream business for over six years. The ten major mid-stream operators are Wide Shine Terminals, Floata Consolidation, Sakoma Midstream Holding (which later merged with Wide Shine), Hoi Kong Terminals, Faith and Safe, Ocean Crown, Fat Kee Stevedores, Singamas, China Merchant, and China Resources. The first three are owned by HIT and control 35% to 40% of the total market share. The next two, owned by Jardine and SHK Properties, enjoy about 20% of the market share. Ocean Crown is owned by Swire. Fat Kee and Singamas are independent, and the last two are China-controlled companies. In addition to these major operators, there are another ten to twenty small, independent operators.

A constraint on the growth of mid-stream throughput is insufficient backup land to store containers before they are picked up by trucks or loaded onto ships. The Public Cargo Working Areas (PCWAs) are often close to busy urban areas and are to some degree controlled by local gangs, so their operation has been rather inefficient. To overcome the limitation of PCWAs, the government has set aside 6.7 hectares of land on Stonecutters Island on which to build a permanent site that is expected to become available in 1998 for use by mid-stream operators.

River Trade

River-trade vessels bring their cargoes to the terminals in Kwai Chung, mid-stream berths, and PCWAs or jetties. Some are serviced at anchorages in the harbour. Small river crafts typically carry anywhere from seven to twenty containers. Since 1992 river trade has referred to the movement of cargoes between the port of Hong Kong and the Pearl River, as well as between the port of Hong Kong and inland waterways in Guangdong and Guangxi Provinces.

To some degree, river trade can be regarded as a response to inadequate road transport between Hong Kong and its neighbouring areas, in particular the Pearl River Delta. Clearly, for cargoes that are near the waterways, river trade is more cost effective than overland hauling, because a river craft can move many more containers than can a tractor-trailer. By providing an alternative to hauling containers to Kwai Chung overland, river trade increases the cargo sources for the terminals in Kwai Chung while at the same time reducing pressure on the territory's road system. However, the fact that many containers are still hauled by tractor-trailers, which might have to wait for hours at the border crossing, indicates that for many shippers, moving cargoes overland is still the preferred mode of transport.

Unlike mid-stream operations, whose container throughput stagnated in 1994 and 1995, the volume of river trade has grown by leaps and bounds. Since 1992, when more reliable data were collected using the current definition, the average annual growth rate has been over 45%. In 1995 river-trade containers reached 1.36 million TEUs and accounted for close to 11% of the overall container throughput. Even though the future of river trade may be uncertain, some major players in the container handling industry see it as a worthwhile direction of diversification.

The first terminal dedicated to river crafts, to be built in Tuen Mun, will be operated by the River Trade Terminal Limited, a consortium jointly led by SHK Properties and Hutchison.[17] It is expected that 800 metres of quay will be completed by 1999, and the total quay front will be 3000 metres long. A second river-trade terminal is also being planned on the north shore of Lantau Island.

Among the existing terminal operators MTL was the only one that did not show interest in Tuen Mun's river-trade terminal.[18] At some point in the past, the container handling industry agreed that there was no logical reason for such a terminal to exist, because it would serve only as a place for consolidation (including repackaging and redistribution) of cargoes from the Pearl River and other nearby waterways. Since containers unloaded at Tuen Mun must be transported to the main terminals at Kwai Chung or to designated parts of the PCWAs where ocean liners are serviced, some would argue that performing consolidation in China would be more efficient than doing so in Hong Kong.

The consolidation function of a river-trade terminal can be served by ports in the Pearl River Delta, and indeed there are many such ports at which goods are sorted and consolidated into containers before they are brought to the territory. Nevertheless, Hong Kong is a hub whereas ports in China are spokes. The spokes do not enjoy the same degree of concentration of activities and scale economy as the hub does. Moreover, even if the consolidation function is ignored, a place is still needed for the river crafts to discharge or load their freight, unless such a facility is provided at the main terminals.

To a certain degree the clientele of river trade might overlap somewhat with mid-stream operations, as river crafts take containers to shore, to the main terminals in Kwai Chung, and to ships served by mid-stream operators. HIT showed a strong interest in the river-trade terminal, partly because it was in the mid-stream business. According to government sources, mid-stream operators would be allowed to use the Tuen Mun terminal. So in a sense a river-trade terminal may also serve as a site for mid-stream operations.

Given the unprecedented growth in river trade, barge berths and associated working areas are to be incorporated into the design of Terminals 10 to 13. The incorporation of barge berths will not only provide the needed river-trade sites but will also reduce barge traffic across main fairways.[19] As a result of the swap agreement involving Terminal 9, two barge berths have been added to the original plan of four ship berths only.

Substitution and Complementarity

In terms of the distance travelled, river-trade cargoes cover the shortest distances, to be followed by intra-Asia cargoes, while trans-ocean cargoes cover the longest distances.[20] Given the cargoes' different characteristics, shipping lines are willing to pay for differential service qualities in terms of turn-around time, storage options and deadline for accepting cargoes. Thus, the three groups of container handling operators — terminal, mid-stream, river — offer services that are at best imperfect substitutes.

As an alternative to road transport, river crafts are complementary to mid-stream operations and terminals in the sense that the former help to bring containers to the territory for handling by the latter. Construction of a river-trade terminal is thus beneficial to terminal operators and mid-stream operators.

Mid-stream operators offer a low-cost alternative to the main terminals. Their greatest appeal is to smaller ships carrying intra-Asia cargoes. Less appealing, but still technically feasible, are the mid-size, ocean-going ships, including intra-Asia and trans-ocean transshipment. Due to the technical limitation of mid-stream operations, large, ocean-going vessels are beyond their reach.

From the perspective of terminal operators as a group, if there are no competing ports in the region, then large, ocean-going container ships are "captured customers", since the competition posed by mid-stream operators is only in servicing small and medium-sized ships.

In reality, however, the port of Hong Kong is also competing against other ports in the region. For transshipment, Hong Kong's operators (be they terminal operators or mid-stream operators) compete against major regional hubs such as Kaohsiung in Taiwan and potential new ports such as Yantian in Shenzhen, China. For direct shipments whose origin or destination is South China, deep-water ports such as Yantian may develop into serious competitors.

In the next chapter, we shall examine in greater depth the structure of Hong Kong's container handling industry, its relationship to the shipping industry, behaviour of firms in each market segment, and different measures of economic performance. The question of competition posed by regional ports will be addressed in Chapter 5.

Notes

1. Based on Tables 2.1 to 2.4, we see that in 1994 the total volume of break bulk was 15,255,000 tonnes whereas that of containerized cargoes was 76,946,000 tonnes.
2. Taylor (1991, p. 109).
3. Sinclair (1992, p. 69).
4. The shares of OCL, Ben Line, and Hapag-Lloyd were 30%, 15%, and 15%, respectively. Other shareholders of MTL included HKBC, Swire Group, and Hutchison
5. See Rodwell (1989, p. 34).
6. Rodwell (1989, p. 48).
7. 25% of Terminal 2's equity was held by Orient Overseas Container (Holding).
8. Sinclair (1992, p. 135)
9. Two ships can be berthed along Terminal 5's 472 metres of quay.
10. ACT's shareholders are Sea-Land (29.5%), Sun Hung Kai Properties (28.5%), Hong Kong Land (22.8%), New World Infrastructure (13.5%), and Jardine Pacific (5.7%).
11. See *South China Morning Post* and *Hong Kong Economic Journal*, 20 September 1996.
12. *Mid-stream Cargo Handling*, PDB leaflet.
13. Port Development Board, *Annual Report 1992*, p. 16.
14. *Mid-stream Cargo Handling*, PDB leaflet.
15. According to Westlake (1991), mid-stream handling was 40% to 50%

cheaper than the terminals were.

16. The government would grant lease of the waterfront up to a maximum of six months, but renewal of lease is possible.

17. The River Trade Terminal Limited is comprised of HIT, SHK Property, COSCO Pacific, and Jardine Pacific. It beat the only other competing consortium, Odelon (which was comprised of New World Infrastructure, SHK Industries, Sealand, and Henderson Land) despite the former's bid ($1.14 billion), which was lower than the latter's ($2.38 billion). According to the government, the HIT-led consortium proposed a better plan to minimize impact on the road system and the environment and to avoid traffic at Mawan.

18. As indicated in the previous footnote, SLOT joined the consortium Odelon to bid for the Tuen Mun river-trade terminal.

19. See PDB's *Annual Report for the period to September 1995*.

20. According to information provided by some informed experts in the industry, transshipment cargoes accounted for 22% and 27.3% of the containers handled by the terminals in 1994 and 1995, respectively. The shares of intra-Asia cargoes for the two years were 36% and 20%, respectively, and those for trans-ocean cargoes were 42% and 52.7%, respectively.

CHAPTER 4

Industry Structure, Firm Conduct, and Economic Performance

Industry Structure

As shown in the previous chapter, container terminals have handled the largest share of the containers passing through the port of Hong Kong, followed first by mid-stream operators and finally by river-trade operators. These three segments of the container handling industry offer different services, charge different fees, and deal with different shipping lines and cargoes.

River-trade operators, by hauling containers between Hong Kong and the feeder ports in the Pearl River Delta, Macau, and ports in Guangxi, are generally complementary to container terminals and mid-stream operators, even though there is some overlap between the services provided by river-trade operators and those by mid-stream operators.

As a low-cost but low-quality alternative to using container terminals, mid-stream operation presents the most serious challenge to the terminals with respect to small ships (several hundred TEUs) and ships carrying intra-Asian cargoes. According to industry experts, ships that carry more than 1,500 TEUs are usually beyond the reach of mid-stream operators. So the terminals face no competition from mid-stream in the case of large ocean liners, whereas mid-stream operators enjoy a clear advantage in the case of small ships. Those in between are within the sphere of competition between the two, and the distribution of cargoes between them depends critically on their relative prices and their handling capacities.

According to an industry analyst 22% of the containers handled by the terminals in 1994 were transshipment cargoes, and 35.5% were intra-Asia cargoes. Both of these were carried by small and medium-sized ships and were thus subject to competition from mid-stream operators as well as from other ports along the China coast and in the region.[1] The remaining 42.5% of throughput were trans-ocean cargoes carried by large, ocean-going vessels, and hence were beyond the reach of mid-stream operators. The shares of transshipment, intra-Asia, and trans-ocean cargoes in 1995 have been estimated at 27.3%, 18% to 22%, and 50.7% to 54.7%, respectively.

Terminal Operators

Among the four terminal operators in Kwai Chung, Hong Kong International Terminals (HIT) is the largest player with or without COSCO-HIT. In 1995 its throughput amounted to 48% of the terminals' total throughput, and its combined throughput with the latter amounted to 62.4% (see Table 4.3 below). In 1993, before COSCO-HIT started operation, HIT's share of the terminals' throughput was slightly over 56%.

Sea-Land Orient Terminals (SLOT) is vertically integrated with its parent company, Sea-Land. While it does move cargoes primarily for Sea-Land, about 35% of its business comes from other shipping lines. In a similar way, COSCO-HIT handles cargoes for shipping lines other than COSCO, but COSCO's business is itself substantial.

Given that SLOT occupies about 12% of the market share and is vertically integrated with Sea-Land, while COSCO-HIT enjoys 14.4% of the market share and largely handles COSCO cargoes, the main competition in the industry is primarily between HIT and Modern Terminals Limited (MTL).

When Terminal 9 is completed there will be a new container terminal operator called Asia Container Terminal (ACT). Since SLOT will be managing the new company's operation, the number of independent operators will not increase, but there will be a more powerful third competitor besides HIT and MTL.

Mid-Stream Operators

Mid-stream operations are not quite capital intensive as the terminals. There are ten major operators and another ten to twenty smaller operators in Hong Kong. Collectively the mid-stream operators form a "competitive fringe", which stands ready to take business away from the terminals if the latter's tariffs are high, or if the latter run up against their capacity limits. This fringe provides a low-price but low-quality alternative to services provided by the terminal operators. It is competitive because entry is easy.

Experts in the industry predict that an increasing amount of the container throughput will come from intra-Asia cargoes carried by relatively small ships. If that is true, then the demand for services rendered by terminals will decline in relative terms, whereas the demand for mid-stream services will continue to grow. However, to the extent that larger vessels are adopted when the volume of intra-Asia trade expands, the relative demand for mid-stream services might decline, unless mid-stream operators can handle increasingly larger ships.

River-Trade Operators

Since the new definition of river trade was used in 1992, the growth of river-trade throughput has been very strong. The average annual growth rate from 1992 to 1995 was almost 47%. Contrary to the very substantial decline in the growth rate of terminal and mid-stream throughput in the first half of 1996, river-trade throughput increased by a remarkable 30% over the first six months in 1995.[2]

Due to low entry costs, there are many river-trade operators, and the business is very competitive. However, competition may decline with the completion of the Tuen Mun river-trade terminal because the first terminal dedicated to river trade will be built by a consortium jointly led by SHK Properties and Hutchison; the market structure could fundamentally change.

Competition between Terminal and Mid-Stream Operators

In economic analysis, the term "oligopoly" refers to a market or industry in which there is a small number of major competitors. Although the term might carry some negative connotations, economists use it in a neutral way to describe an industry with a small number of competitors. The term "competitive fringe" refers to a large number of small firms operating competitively (i.e., no single firm has any visible market power due to small market share and easy entry) along with one or several major firms. From the above discussion, we see that the industry handling ocean-going containers is an example of an oligopoly (few terminal operators) facing a competitive fringe (many mid-stream operators). It is a slight variant of the standard model of an oligopoly with a competitive fringe in the sense that in the container handling industry the fringe offers an imperfect substitute that is of lower quality.

The terms "upstream" and "downstream" are used in economic analysis to describe two industries that are vertically related in a multi-stage production process. The output of an upstream industry is used as an "intermediate input" by a down- stream industry whose output is used either by the final consumers or by an industry that is further down the "stream". None of the above should be confused with the term "mid-stream" here which refers to a particular way of handling port cargoes.

The container handling industry is an interesting example of vertical relations between container handlers in the "upstream" and container carriers in the "downstream". Shipping lines compete for cargoes from numerous shippers (who include manufacturers, traders and retailers). The quality of container handling in the upstream affects the quality and cost of services provision by shipping lines in the downstream. There are about two dozen or so major shipping lines, so they are neither perfectly competitive (a very large number of competitors) nor oligopolistic (a very small number of competitors who are keenly aware of their interdependence). In economic terms, these shipping lines may be regarded as "monopolistically competitive". But since the shipping industry is currently being grouped into five or six consortia, it would be more appropriately regarded as an oligopoly.[3]

In either case, the "container handling" and "container shipping" industries are an example of a two-level oligopoly, where there are more firms in the downstream than there are in the upstream. With the emergence of shipping-line consortia, the difference in upstream and down-stream numbers has declined.

The difference in market structure of the upstream and downstream industries means that the downstream firms have relatively weaker market power than do the upstream firms, especially when the former suffer from serious excess capacity.The existence of a competitive fringe in the upstream, however, serves to impose a limit on the extent to which the upstream oligopolists (terminal operators) can exercise their monopoly power.

It would be difficult to find clear signs of the ebbs and flows of mid-stream throughput relative to the capacity of the terminals, because the composition of different types of cargoes (intra-Asia versus trans-ocean and transshipment versus direct shipment) have changed over time. However, the periods of slow growth in mid-stream container throughput did correspond to the addition of new berths in Kwai Chung. For instance, as is illustrated in Table 3.20, there were two periods of slow growth in mid-stream throughput, 1989-1990 and 1993-1995. The first slow period coincided with the completion of Terminal 6 in 1989. The second coincided with the completion of new berths on Terminal 8 beginning in 1993, the provision of additional backup land, and capacity-enhancement investment, have all led to an increase in the terminals' handling capacity.

Horizontal Integration by Terminal Operators

Some container terminal operators have integrated horizontally into the mid-stream and river-trade business. For instance, three major mid-stream operators accounting for 35% to 40% of the market share are owned by Hutchison, a subsidiary of which, HIT, is the largest terminal operator. Two major mid-stream operators accounting for about 20% of the market share are owned by Jardine and SHK Properties, both being the shareholders of ACT. Finally, the Tuen Mun river-trade terminal will be built and operated by a consortium comprised of HWL, SHK Properties, COSCO-Pacific and Jardine.

In contrast, MTL has not shown much interest in either mid-stream or river trade; it focuses only on trans-ocean containers. SLOT tried to move into river trade, but the attempt was not successful.

One might wonder whether the above horizontal integration by some terminal operators could have increased their monopoly power and enabled them to set higher terminal tariffs. Would the interrelated ownership cause a problem, and would it be desirable to disallow horizontal integration? The answers to these questions depend on whether the mid-stream operators act independently of, and put competitive pressure on, the terminal operators, as one would expect from a competitive fringe. A key consideration is whether there are barriers of entry into the mid-stream business. Given that entry costs are low and entry is relatively easy, ownership of mid-stream operators by terminal operators is unlikely to be consequential. The reasons are (a) the mid-stream operators unrelated to the terminal operators are ready to expand, and (b) potential entrants are ready to enter the market should "tariffs" (i.e., fees charged by terminal operators for container handling) become abnormally high.[4] In the case of river trade, physically small river crafts are used, so entry costs appear to be even lower so long as there are areas available for the loading and unloading of containers. More importantly, river trade is complementary to container handling both mid-stream and at the terminals.

In summary, ownership links represent a market solution

whereby efficiency can be enhanced via joint ownership of more than a single type of operation. It is a conscious choice made by the operators depending on their chosen spheres of business. For instance, MTL has chosen to concentrate only on the handling of trans-ocean containers, but Hutchison and its subsidiary HIT has chosen to diversify into all three modes of container handling.

Conduct of the Firms

Oligopolistic Interdependence

Terminal operators all claim that they engage in fierce competition with one another for customers. However, there are simply too few players for them not to have any explicit or implicit understanding about the avoidance of head-on competition. It would be unbelievable indeed if they failed to recognize their close interdependence and to adjust their competitive strategies accordingly. The upshot is that competition is less intense than it would be if the industry were comprised of many more terminal operators. Some operators do admit that such an inference is not far from the truth.

Given the structure of the shipping industry, each terminal operator has a relatively small number of major customers. Terminal operators typically negotiate a separate contract with each customer (i.e., a shipping line), so there is room for price discrimination (i.e., preferential treatment of some clients). Due to differences in services provided — priority berthing, free storage periods, exit conditions — it is difficult to compare tariffs. The premium for most service attributes is said to be fairly standard in the industry, so the most important instrument of price discrimination is the degree of volume discount. Unfortunately, information about tariffs is a trade secret. We have therefore been unable to obtain information even about the degree of volume discount.

Typically, contracts are three to five years in length, but there is a tendency towards shorter contracts. Being conscious of their interdependence, terminal operators seldom try to steal existing customers from competitors for fear of upsetting the equilibrium, except when changes occur among the consortia. Indeed, according to industry insiders, operators have "non-poaching" agreements that could be prohibited by "fair trade" and "anti-trust" laws. Nonetheless, they maintain contact with their competitors' customers, because the latter are future potential clients. In case re-alignment of shipping lines occurs, the affected shipping lines will be regarded as "new business" that is up for grabs.

Terminal operators also co-operate in other ways. For instance, they may loan berths to one another to deal with unusual bunching of ship arrivals for commercial reasons, although in practice the arrangement is infrequent except for SLOT's routine use of HIT's Terminal 4. They discuss things that are of common interest. They obtain information from each other and quite often via the Port Development Board (PDB) and the Marine Department.

Thus, there seems to be some evidence that the existing container terminal operators have engaged in interactions that are not unexpected of oligopolists. Having only two major independent operators makes it much easier to enforce any explicit or implicit agreements about tariffs and competition. Such co-operation would naturally lead to higher tariffs and larger profits. Thus, the validity of complaints about the terminal operators' virtual monopoly pricing, though hard to substantiate without the confidential data about tariffs and unit costs, would not come as a surprise under the circumstances. (Some information about tariffs will be provided in Table 4.2.)

Although HIT and, to a lesser extent, MTL do own and operate ports in South China, these ventures seem to be more motivated by new business opportunities rather than simply by a desire to influence (shore up) tariffs charged at Hong Kong terminals. If anything, aggressive development of competing ports in the region would weaken terminal owners' competitive position in Hong Kong. HIT and MTL are, however, seeking to improve their overall position in the region.

Relationship with Shipping Lines

A recent trend in the shipping industry is for major shipping lines to form alliances or consortia to improve their services and bargaining strength while holding down costs. The shipping alliances which are most important to Hong Kong's cargo movement are (1) Maersk and Sealand, (2) the Global Alliance which has five members,[5] (3) the Grand Alliance which has four members,[6] and (4) Hanjin, Cho Yang and DSR-Senator.[7] In addition, Evergreen and COSCO are also very important. Each alliance negotiates with the terminals to obtain identical terms for its members.

While the formation of the above alliances was motivated by the shipping lines' need to pool resources to improve their services to the shippers while holding costs down, the alliances seem to have put pressure on Hong Kong's terminal operators to offer better services and lower tariffs. As an indication of the alliances' bargaining power, a single alliance might be able to shift enough cargoes (say, more than 1 million TEUs) to make a new port viable. To the extent that an alliance makes more ship calls to the same port than does any single shipping line, the formation of alliances tends to favour operators with more berths. Thus, smaller operators, who are unable to service large consortia volumes, are keener than are larger operators to increase capacity. As we shall see in Chapter 8, this prediction is borne out by the operators' actual positions on new terminals.

Before the formation of alliances, dedicated berths were an important tool of competition among shipping lines. In Hong Kong, however, common user terminals are the norm, due to the lack of land. With alliances, berths dedicated to specific shipping lines are not only unnecessary, but could also lead to wasteful "double berthing".

Perspective of the Shippers

The Shippers' Council is an association of importers, exporters and manufacturers whose goals are to work for low shipping rates and good shipping services. Since most Hong Kong manufacturers export their products under FOB (free on board) terms of sale, they are responsible for the "terminal-handling charges" (THCs) which are fees charged by shipping lines for loading their cargoes into the ships designated by their overseas buyers. In the rest of this book, we maintain a distinction between "tariffs" (terminal fees paid by shipping lines to terminal operators) and THCs (terminal fees paid by shippers to shipping lines).

According to the Shippers' Council, the assessment of THCs began in 1990 with the Far Eastern Freight Conference (FEFC). For the first time, "shore side charges" were separated from "freight rates". Later, the Intra-Asia Discussion Agreement (IADA) and Asia North America Eastbound Rate Agreement (ANERA) followed suit.

The proposal to separate THCs from freight rates was regarded by shippers at the time as an improvement over the previous arrangement, because the separation was meant to increase the transparency in the setting of shipping rates. However, shippers were disappointed by the subsequent implementation of the proposal. First, THCs were supposed to recoup about 80% of the tariffs paid by shipping lines to the terminal operators, but THCs in Hong Kong rose at an average rate of over 10% per annum, in contrast with almost constant THCs in North America and Europe.

Informed sources in shipping lines explained that the relatively rapid increases in THCs were largely a result of the fact that at the beginning only a small fraction of the actual terminal tariffs was passed on to the shippers in the form of THCs. The fraction was raised gradually over time, thus resulting in a higher rate of increase in THCs than in terminal tariffs. This practice led to the erroneous belief that the terminal operators had increased their charges substantially over time. Regardless of the real reasons for the increases, rapidly rising THCs were especially disturbing to shippers on the FOB term of sale when freight rates were falling.

Second, shippers found it very difficult to recover the increases in THCs from their overseas customers. Manufacturers and exporters had limited ability to raise the FOB price of their products, because they faced stiff competition from suppliers in other parts of the world such as Taiwan, Korea and Southeast Asia. But since the manufacturers knew what the THCs were when they bargained with overseas buyers, they should have incorporated them in their reservation prices. Needless to say, any increase in THCs, other things being equal, would hurt shippers and manufacturers, just as any other cost increases would.

Shipping lines saw the introduction of THCs as a cost-recovery tool. They used THCs to avoid irrational competition by alerting their own sales representatives to the existence of such a "cost" item. As shipping lines expected, the introduction of THCs reduced the extent of excessive competition and improved their profits.

Shippers suspect that shipping lines have used THCs as an instrument with which to squeeze as much profits from them as possible. But given that (a) the overseas buyer and the local manufacturer-exporter together bear the sum of the THCs and freight rate, and (b) the manufacturer-exporter knows that he is responsible for the known THCs, a question arises: would the equilibrium outcome of charging a freight rate that is inclusive of THCs be any different from that of charging THCs and freight rates separately? Could the situation be similar to the incidence of a sales tax? In the case of a sales tax, how much of the tax is borne by consumers and how much is borne by producers depends on the relative price elasticities of demand and supply, but it is independent of whether the tax is collected from the consumers or from the producers.

To answer the question of whether the separation of THCs from freight rates might increase shipping lines' profit, we have analyzed a model of two shipping lines (the analysis can be generalized to any finite number of shipping lines) engaging in standard oligopolistic competition,[8] subject to the buyers' demand function and the manufacturers' supply function. The total volume of cargoes depends inversely on the sum of a THC and a freight rate.

In our analysis, we assume that the two shipping lines are able to set a common THC that is binding on both of them, but that they cannot set any common freight rate, so that they compete with each other in freight rates given whatever level of THC. These assumptions approximate the relative degree of binding power of the THCs and freight rates set by various international conferences. For instance, IADA does not publish its freight rates, which are intended only as suggestions. FEFC sets freight rates but cannot enforce them. The freight rates set by ANERA must be followed, at least on paper, but shipping lines can still cheat by offering discounts with through bills. In contrast, even though small shipping lines adjust their THCs in response to supply and demand conditions and client attributes, large shipping lines typically stick with the THCs set by the conferences.

The above assumption about the relative binding power of THC and freight rates implies that we should analyze a "two-stage game" in which a common THC is set in the first stage, and the shipping lines engage in standard oligopoly competition in the second stage given the chosen THC. In this game, it can be proved that an increase in THC will reduce the freight rate, but less than one for one, thus resulting in an increase in the sum of the two. The result holds in general, so that it covers the special case in which the THC is set exactly equal to the terminal tariff purely as a cost-recovery device.

Starting from no THC, a small THC will necessarily be profit increasing for the shipping lines. The intuition for this is rather straightforward and can be interpreted as a special case of a general result established in oligopoly theory. We know that the combined output of two oligopolistic firms is larger than the output of a single monopolistic firm, because competition is at a minimum when there is only a single monopolistic firm. An increase in THC reduces each shipping line's incentive to supply shipping services and hence reduces the extent of competition. Unlike an indirect tax assessed by a government, the THC is part of the shipping lines' revenue.As a result, any small increase in the total price (THC plus freight), by reducing supply, must increase their total profits.

If shipping lines have identical cost conditions, then any increase in total profits must also result in a corresponding increase in each shipping line's profits. Indeed, one can further show that there exists some THC that will reproduce the industry's monopoly solution.[9] In essence, the THC can serve as a mechanism to induce a monopolistic outcome in an oligopolistic setting.Thus, even if shipping lines' sales representatives are rational in the sense of taking the tariffs into account in their own pricing decisions, the existence of an appropriate THC will still contribute to profits.

We are not suggesting that shipping lines use THCs as a price tool to reduce competition and increase profits. What we have shown is that under certain realistic assumptions THCs can be used for such purposes. Thus, the shippers' suspicion is not without grounds.

Shippers feel that THCs are high due partly to high terminal tariffs, a result of insufficient competition with few independent operators and berths. They would prefer to see more competitions among terminal operators, and perhaps government regulation.

In the case of shipping lines, shippers have mixed feelings about the formation of shipping consortia, which increase shipping lines' bargaining power vis-a-vis not only terminal operators, but also vis-a-vis shippers themselves. If the shipping consortia were to continue, shippers would like to require shipping lines to reveal information about their costs, including the tariffs paid by shipping lines to the terminals. Since shipping lines operate across national borders, such a proposal would require the Hong Kong government to enter into negotiation with foreign governments.

The quality difference between container handling at the terminals and that at mid-stream is of little concern to shippers who operate on FOB terms of sale, because their tasks are completed as soon as their cargoes are loaded into the designated ships. Thus, the concern about quality differences is mainly a matter for shipping lines, which care about the cost of turn-around time, and shippers and manufactures who operate on CIF (cost, insurance, freight) terms of sale. In a similar vein, shippers are more concerned about inexpensive shipping costs (low THCs and high-quality service) and convenience, but not about the exact location of berths. Since Hong Kong has sufficient throughput to feed its terminals, they believe there are enough additional cargoes to feed other ports in the region.

Entry Deterrence and Pre-emption

One might ask whether the major players in the container terminal business could prevent entry or could engage in predatory pricing to drive out new entrants. Given that the government decides on whether new terminals should be built and on the method for selecting which companies are to be awarded the right of building and operating these terminals, the existing terminal operators might attempt to influence the government to their advantage on both decisions. As we shall see in Chapter 8, existing large operators and small new entrants hold diametrically different positions with regard to the need for additional terminals on Lantau.

Naturally, existing operators would attempt to find ways to protect their past investment. One would expect them to engage in some forms of pre-emptive bidding for new terminals to prevent new entry that could cause "dangerous competition". Operators currently make no secret of this concern. Sinclair (1992, p. 73) describes MTL's past concern about the award of Terminal 5 as follows, "If MTL did not build more berths, someone else would .... They were reluctant to commit their scarce resources. Yet to stand back whilst others built the berth would create dangerous competition."

In principle, existing operators can out-bid new entrants if new entrants dilute total industry profits. This is a likely outcome, because additional competitors would make cartel agreements more difficult to enforce. However, the government is not necessarily bound to award development rights to the higher bidder, particularly if it sees the need to increase the degree of competition or to bring about a more acceptable division of profits. This view is expressed most explicitly by the Hong Kong Centre for Economic Research (1992).

If new entry cannot be blocked by persuasion or pre-emptive bidding, can existing operators engage in predatory pricing and cut-throat competition to drive out a new entrant? What is the minimum number of berths necessary for a new entrant to enjoy the economy of scale in order to be viable?

The answer to the last question seems relatively straightforward. Although some in the industry feel that it is not economical to operate a single berth, the fact that SLOT has operated a single berth profitably indicates that the minimum scale of efficient operation is not very large. Industry experts suggest that one million TEUs would be sufficient to exploit the economy of scale. Thus, having two berths for ACT should be sufficient.

More important is whether a new entrant has to incur the fixed cost to develop its operating and information-management systems. If ACT must do so, then perhaps there is cause for concern. However, the operator of ACT is SLOT, so it is not necessary to spend a very substantial amount of resources to re-invent the wheel. Moreover, by several measures SLOT is apparently the most efficient operator in Kwai Chung. Thus, predatory pricing by others will have little chance of success.

The operating systems appropriate for Hong Kong are not found in other container ports in the world, because containers are stacked up higher in Hong Kong than at other ports due to limited backup land. Thus, new entrants might not have a strong incentive to enter the terminal business, unless they can align themselves with one of the existing operators or have the latter manage the new terminals for them. If the government really wants new operators, it can accept a lower bid from an outside bidder, and the difference between that and the highest bid from an existing operator can be regarded as a subsidy for the initial set-up costs.

In sum, given the importance of industry-specific and even firm-specific capital in the form of operations and information management systems, building new terminals might not necessarily lead to a larger number of independent operators if the award of terminals is determined purely by the size of bids.

Table 4.1
A Comparison of Terminal Handling Charges (THCs)
(Exchange Rates as of 10 October 1996)

	Transpacific Eastbound ANERA		Transpacific Westbound TWRA		Asia/Europe FEFC		Intra Asia IADA	
HK$	TEU	FEU	TEU	FEU	TEU	FEU	TEU	FEU
Hong Kong	1,875	2,500	1,875	2,500	1,686	2,491	1,380	2,070
Taiwan	976	1,440	1,224	2,054	1,224	1,580	1,224	1,580
Singapore	996	1,478	1,297	1,921	887	1,259	996	1,478
South Korea	770	1,105	818	1,107	818	1,107	651	958
Japan*	Free	Free	Free	Free	Free	Free	763	1,145
Malaysia	876	1,306	1,177	1,758	584	876	907	1,352
Philippines	541	734	773	958	502	618	434	578
Indonesia	657	1,082	N.A.	N.A.	N/A	N/A	387	580
Thailand	798	1,197	798	1,197	N/A	N/A	798	1,197
Germany	N.A.	N.A.	N.A.	N.A.	1,504	1,504	N.A.	N.A.
Netherlands	N.A.	N.A.	N.A.	N.A.	1,364	1,364	N.A.	N.A.
U.K.	N.A.	N.A.	N.A.	N.A.	831	831	N.A.	N.A.

ANERA = Asia North America Eastbound Rate Agreement
TWRA = Transpacific Westbound Rate Agreement
FEFC = Far Eastern Freight Conference
IADA = Intra-Asia Discussion Agreement
TEU = Twenty-foot Equivalent Units
FEU = Forty-foot Equivalent Units
N.A. = Not Applicable
N/A = Not Available

* Freights between Japan-North America and Japan-Europe are under different conferences. There are no THCs charged on Japanese exports to North America and Europe. Similar charges imposed by the IADA are known as the Empty container handling Charge (ECHC).

Table 4.2
Tariff, CPI(A), and Unit Operating Cost

	1991	1992	1993	1994	1995
Tariff	100	109.3	115.7	123.5	129.1
	—	(9.3%)	(5.9%)	(6.7%)	(4.5%)
CPI(A)	100	109.4	118.8	128.3	139.5
	—	(9.4%)	(8.5%)	(8.1%)	(8.7%)
Unit Operating Cost	100	96.8	94.0	88.2	87.0
	—	(– 3.2%)	(– 2.9%)	(– 6.2%)	(– 1.4%)

Economic Performance

THCs: Fees Charged by Shipping Lines

Hong Kong's THCs are known to be among the highest in the world. A comparison of its THCs with those of other East Asian ports as well as with those of Germany, the Netherlands and the U.K. as supplied by the Shippers' Council is reproduced with some slight modifications in Table 4.1. As the table shows, Hong Kong's THCs are not only the highest but exceed those by the other ports by a large margin. The level of THCs is higher in Hong Kong because government subsidies are absent, unlike other ports such as Singapore.

Tariffs: Fees Charged by Terminals

Shippers often complain about high THCs and attribute them to the monopoly power (or oligopoly power) of terminal operators. In response, the terminal operators waste no time in pointing out that (a) THCs are charged by shipping lines,[10] not by container terminals, and (b) changes in THCs do not necessarily correspond to changes in tariffs, the container handling fees paid by shipping lines to the terminal operators. Shippers claim that the THCs they pay the shipping lines make up an increasingly larger percentage of the tariffs, thus giving rise to a misconception that terminal operators cause the THCs to rise. Terminal operators further point out that in the last seven years the increases in tariffs were on average three percentage points below the rate of inflation, implying that in real terms the tariffs have been declining over time.

Table 4.2 is based on information provided by an industry analyst, including the indices and percentage changes of the representative tariffs, consumer price index (CPI(A)), and unit operating costs of a major terminal operator during the period 1991–1995. The percentage changes are in parentheses. Between 1991 and 1995 the tariff rose at an average rate of 6.6% per annum, which was below the average rate of inflation at 8.7%. However, the unit operating costs declined by about 3.4% per annum. That is to say, the ratio of tariff and unit operating cost has been increasing at the rate of 10% per annum. Since finance charges and depreciation are excluded from the operating costs, we have no information about the relationship between tariff and total cost per unit. But unless the two omitted items have risen at a very fast rate (e.g., exceeding 10%), the profit margin has been widening.

In addition to shippers, who complain about high shipping costs, shipping lines also complain about high tariffs as a result of monopoly pricing.[11] The real reason for the shipping lines' complaints, argue some terminal operators, is excess capacity in the shipping industry. These terminal operators also believe that due to the bad financial situation of the shipping lines, a reduction in tariffs by terminal operators would not result in any savings to the shippers. That being the case, a reduction in tariffs would not lead to any significant increase in the volume of throughput, and hence it is not profitable to reduce them.

It would be hard to deny, and some terminal operators do admit, that with their small number, the incentive to cut prices to compete for customers is not as strong as it would be if there were many competitors. In their defence, the PDB points out that, with four or five terminal operators plus mid-stream operators, there is more competition in Hong Kong for container handling than there is in any other port in the world. In addition Hong Kong also faces intensive competition from other ports in the region.

Regardless of the connection between high tariffs and monopoly power, most would agree that high costs are one reason for high tariffs. Compared with its regional competitors such as South China, Taiwan and Southeast Asia, the costs of labour and land in Hong Kong are relatively high.[12]

How do Hong Kong's tariffs compare with tariffs in other ports in Asia? Many claim that they are among the highest, but some operators also point out that Japanese tariffs were higher when the yen was strong. Some also argue that Singapore's tariffs would be comparable to Hong Kong's if the subsidies provided by the Singapore government in the form of savings in interest, land premium, no dividends, and taxes were added back. Finally, the PDB points out that Hong Kong's higher tariffs should not be confused with the total costs of using the port of Hong Kong, which are comparable to those for using most ports in Asia, because other port charges are lower in Hong Kong. According to Scofield and Boyce (1991), the average tariff for handling a TEU in 1991 was HK$1,000. The figure for the first half of 1996 was about HK$1,300 per TEU.[13]

It is natural for shippers and shipping lines to complain about terminal operators' tariffs. As a director of the Hong Kong Shipowners' Association put it, "shipowners will always say port charges are too high, but that's business .... You get what you pay for. If it really was too expensive, they would go elsewhere."[14]

Returns on Investment

An indirect way of approaching the issue of monopoly pricing is to examine the terminal operators' profitability and rate of return on past investment. Unfortunately, these financial data are confidential information not available to the public. Informed sources in the industry, however, have put the rate of return on capital investment at 16% to 17%, which appears to be a very good rate of return, but perhaps that rate is not out of the ordinary when compared with the rate of return of major companies, including public utilities in Hong Kong.[15]

Historically, HIT made handsome profits for its parent company. According to Rodwell (1989, p. 51), in 1977, HIT "represented only 8% of the (Hutchison Whampoa Limited) group's turnover, its contribution to group profit was no less than 24%." In 1988 "HIT's contribution to the fortunes of its parent company remained high: while the turnover of these companies represented 9% of group turnover their profit contribution was almost 29%, most of which was attributable to container handling." (p. 61)

The sketchy information about returns to investment and the relationship between tariff and unit operating costs is not inconsistent with the general impression that container terminals are in a very profitable and stable business.

Table 4.3
Market Shares of Kwai Chung's Terminal Operators

	1991	1992	1993	1994	1995
MTL	34.6%	33.0%	30.5%	27.3%	25.6%
HIT	51.3%	54.5%	56.1%	52.2%	48.0%
SLOT	14.1%	12.5%	13.4%	11.8%	12.0%
COSCO-HIT	—	—	—	8.7%	14.4%
TOTAL	100.0%	100.0%	100.0%	100.0%	100.0%

Table 4.4
Productivity (TEUs) of Terminal Operators in 1995

	Per Berth	Per Metre of Berth	Per Hectare of Land
MTL	422,300	1,159	24,800
HIT	396,400	1,204	44,500
SLOT	880,300	2,886	52,700*
COSCO-HIT	596,500	1,864	39,800

* See endnote 16 for elaboration.

Market Shares and Productivity

Let us examine the market shares of the existing terminal operators from 1991 to 1995, keeping in mind that SLOT was set up in 1991.

From Table 4.3, we see that both SLOT and MTL lost market shares to HIT and COSCO-HIT during this period. With only one berth plus some help from HIT's Terminal 4, it was quite remarkable for SLOT to achieve a share of 12% in 1995, in contrast to MTL's continuous downward slide from 34.6% in 1991 to 25.6% in 1995. On the other side of the same coin, HIT and COSCO-HIT managed to increase their combined share from 51.3% in 1991 to 62.4% in 1995. HIT's own share, however, fell as COSCO-HIT grew.

Since the above four operators had different facilities at their disposal, it is helpful to examine their productivity with different productivity measures (per berth, per metre of berth, and per hectare of land). We hope that collectively these data provide us with a more complete picture of the operators' relative efficiency. In our calculation of SLOT's productivity, we have used the throughput of Terminal 3 only; throughput handled by SLOT using HIT's Terminal 4 has been excluded.

In Table 4.4, the productivity figures for 1995 per berth and per hectare of land are rounded to the nearest hundred TEUs. The figures show that by any of the three measures, SLOT was most efficient. For the rest, COSCO-HIT had high productivity per berth and per metre of berth, whereas HIT had high productivity per hectare. MTL came in last except in productivity per berth.

It is interesting to note that in each of the three different dimensions, SLOT was not only more productive than were the two largest operators, but it was more productive by a large margin.[16] One factor behind SLOT's superior performance was the vertical integration it has achieved with its parent, Sea-Land shipping line. On the one hand, to the extent that SLOT has to provide additional services to meet the needs of Sea-Land, the vertical relationship with its parent might impose constraints on its own flexibility. On the other hand, the relationship ensures a stable demand for its container handling services. SLOT has used HIT's Terminal 4 to handle some of its containers, because demand has exceeded its own operating capacity at Terminal 3. This excess demand might have given SLOT an incentive, unmatched by other terminal operators, to enhance the productivity of Terminal 3. As a result of the increase in productivity, the throughput handled by SLOT at its own terminal has continued to rise, but the throughput by using HIT's Terminal 4 has declined since 1994.

Another factor might be due to SLOT's competition for efficiency against other Asian container terminals owned by Sea-Land, including those in Kobe, Yokohama, Puhan and Kaohsiung. A key measure of the operating efficiency of container terminals is "port moves per hour" (PMPH). SLOT's current performance of around forty PMPH is in the forefront among Sea-Land container terminals in East Asia. The speed at handling containers has translated into an important competitive edge, namely, quick turn-around time for ships.

Some industry experts explain that a constraining factor on productivity is not the quay, but how much backup land is available for stacking and storing the containers to be loaded and unloaded. A main focus of SLOT's productivity enhancement effort is to streamline and co-ordinate operations in the container yard, including a "pre-stacking" step to provide a desirable configuration of containers before the actual loading of containers into ships or trucks. If operations in the yard are efficient, they will not become a bottleneck for the quayside gantry cranes. Thus, even if backup land is the only critical factor of productivity, there seems to be ample room for HIT, COSCO-HIT, and especially MTL to improve the productivity of their existing backup land.

Overall Performance

By any standard Hong Kong's container port has been very successful in supporting the territory's own exports and imports as well as in handling entrepôt cargoes. The fast-expanding, trade-related industries have contributed greatly to China's expansion in trade with the rest of the world. In the process, they have also significantly increased output and income of Hong Kong. The container handling service providers in the private sector have certainly responded well to demand.

Aggregate Economic Welfare and Monopoly Tariffs

From the point of view of Hong Kong's aggregate economic welfare, terminal operators' charging higher-than-competitive tariffs on entrepôt cargoes and reaping above-normal profits are not necessarily bad. Given Hong Kong's strategic position on the trade routes, such tariffs might just be right. Part of the higher tariffs are ultimately borne by consumers and producers outside Hong Kong, and it would not be in Hong Kong's best interests to charge the foreign firms competitive tariffs that only cover costs. As an analogy, the Organization of Petroleum Exporting Countries

(OPEC) would not want to charge competitive prices for its oil.

One may define "socially optimal" tariffs as tariffs that maximize Hong Kong's aggregate economic welfare and to define monopoly tariffs as tariffs charged if Hong Kong's container terminals were owned by a single company. In Chapter 8, we shall discuss these two tariffs as well as competitive tariffs in greater detail when we try to determine the desirable degree of competition in the industry.

While existing terminal operators are collecting monopolistic rent from their past investment, they are also under increasing pressure from regional competitors, in particular Taiwan and the emerging ports in South China. Regardless of the degree to which these competing terminal ports are owned by the terminal operators in Hong Kong, pressure on the latter will increase. As a result, the monopoly position of Hong Kong's terminal operators would only be temporary, since other ports in the region are becoming more serious competitors. In the next chapter, we shall examine port development in the region, and the various implications for Hong Kong.

Notes

1. According to some shippers, a large portion of the intra-Asia cargoes are handled at the berths.
2. The PDB (1996a) has predicted that river trade throughput would increase at the rate of 27% until 1998, and then at 14.4% per annum until 2001. After that, throughput would continue to grow at 9.8% per annum, to 5.8 million TEUs by 2006.
3. In a report in the *South China Morning Post*, 11 January 1996, the Director of the PDB was quoted as saying that "in 1992, there used to be 40 or so shipping lines. Now there are half a dozen huge shipping consortia."
4. Since mid-stream operators and container berths provide services that are only imperfect substitutes for a certain range of ships, one might argue that controlling supply in the former would have a limited effect on the latter's market power. However, if the argument is valid, then it would also imply that mid-stream does not exert very much competitive pressure on the terminals.

5. The six members of Global Alliance are (1) Mitsui OSK Lines, (2) MISC, (3) Orient Overseas Container Line, (4) American President Line, and (5) Nedlloyd.

6. The four members of Grand Alliance are (1) NYK Line, (2) Neptune Orient Lines, (3) P&O Containers, and (4) Hapag-Lloyd.

7. According to some terminal operators, about 60% of all containers passing through Hong Kong are carried by these four consortia.

8. More precisely, Cournot competition or Bertrand competition, in economics jargon.

9. A mathematical proof of this and the above results are available from the first author upon request.

10. According to certain estimates, about 80% of the tariffs are currently passed directly to shippers as THCs, on which shipping lines do not pay commission to agents. But commission is not a consideration for large shipping lines because they do not use agents. In addition to THCs, shippers also pay freight rates, document fees, bunker adjustments, currency adjustments, and congestion adjustments.

11. According to the *South China Morning Post*, 14 March 1996, the Chairman of the Hong Kong Liner Association claimed that HIT and MTL were "virtual monopolies which charged exorbitant fees for handling containers."

12. According to informed sources, the development cost (construction plus equipment) per berth in the mid-seventies was about $120 million. In contrast, the development cost on Terminals 7 and 8 was about $2 billion per berth. The development cost on Terminal 9 is expected to exceed $3 billion.

13. The increase in tariffs between 1991 and 1996 is not inconsistent with information provided in Table 4.2.

14. *South China Morning Post*, 20 September 1996

15. As summarized by Lam (1996, Table 6.9, p. 115), the regulated rates of return for electricity, bus, and air terminals are 13.5%, 15% to 16%, and 18% of their respective "average net fixed assets". The regulated rate of return for air cargo is 12.5% of the "gross value of assets".

16. According to officially announced figures, SLOT has 16.7 hectares, but if the area in Mei Foo is included then SLOT has a total of 19.1 hectares and the productivity (TEUs) per hectare of land becomes 46,000 which is still the highest, but the margin is much smaller.

CHAPTER 5

Regional Port Development: Competition and Co-operation

Do Ports Follow Cargoes or Vice Versa?

It is hard to give a definite answer to the question of whether ports follow cargoes or vice versa. However, if we see the relationship between ports and cargoes as a process of dynamic evolution, then some more definite statements can be made. First, ports require certain natural conditions (deep water, shelter from waves and so on) and take a long time to develop. Thus, at least in the short term, economic activities tend to concentrate around good ports and other convenient nodes of transportation. As the volume of cargoes rises with the expansion of economic activities, the adequacy and cost effectiveness of using the same port facilities might be questioned. If facilities could be expanded and improved at a reasonable cost, then the existing ports would continue to grow and might even increase their attractiveness through the "economy of agglomeration".

But if serious bottlenecks develop in the supporting infrastructure such as roads and railways that connects a port to cargo sources, then the port's economy of agglomeration might give way to the diseconomy of congestion. And when congestion becomes a long-run feature as opposed to just a short-run phenomenon, shippers and shipping lines would look for alternative ports. Nonetheless before the volume of overflow cargoes from the existing main ports has reached a critical threshold, the new ports would normally be struggling under insufficient scale and the lack of agglomeration economy. Once sufficient cargoes have moved to the new ports, however, a virtuous circle would begin. The new ports would be increasingly attractive and cause a diversion of cargoes beyond overflows from the established hub.

China's economic reform and open-door policy have resulted in very rapid increases in the volume of its foreign trade. A substantial fraction of its trade has been channelled through Hong Kong, primarily through the port. The port of Hong Kong's predominant role in handling China's foreign trade was due as much to China's underdeveloped port facilities as to Hong Kong's strategic location and superior facilities, in particular facilities for handling container ships. Moreover, an increase in the volume of cargoes has also led to the emergence and improvement of supporting services, thus further enhancing Hong Kong's attractiveness as a hub port.

However, as China's foreign trade continued to expand, the high cost of moving cargoes through the port of Hong Kong became an increasingly important issue. The need to develop modern port facilities with high-quality services along the Chinese coast became widely recognized. It is not a question of whether Hong Kong or which other Chinese ports should be handling the cargoes, but a question of how many more ports are needed along with the port of Hong Kong.

Parochial interests should be ignored. It is abundantly clear that a network of ports, appropriately located in different parts of China, will be required to serve the need of regional trade. An expansion of trade volume and a dispersion of cargo sources would enable more ports to enjoy the needed economy of scale; it would also demonstrate the ineffectiveness of moving a disproportionately large share of cargoes through a single port.

The above logic applies more broadly to East Asia also, suggesting that a network of regional ports will emerge in response to the growth and shifts of cargo sources. However, in our discussion of regional competitors, we shall focus only on port that are sufficiently close to Hong Kong so that there is a significant overlap in their cargo sources.

Regional Ports: Competitive or Complementary?

Are the existing and emerging ports in the region competitors of the port of Hong Kong? Or are they complementary to one another? It is not possible to answer this question unambiguously without knowing what functions each port serves. Furthermore, the answer would seem to depend on whether one approaches the question from a local or a regional perspective.

While Hong Kong's deep-water port serves ships of all sizes and types, it handles predominantly ocean cargoes that travel long distances. Thus, from this perspective, all ports in the Pearl River Delta that have no deep-water berths and approach channels are not competitors of, but are rather complementary to, the port of Hong Kong. However, if ports in the region have the capacity to handle ocean cargoes too and can provide more or less similar services, then they are Hong Kong's potential competitors.

From Hong Kong's perspective, its port and the regional ports may be complementary if the emergence of new ports enables Hong Kong to benefit by specializing in a certain subset of cargo-handling activities such that the total value-added of its specialized activities increases. As an analogy, the ability to shift labour-intensive functions of the industrial production process to the Pearl River Delta has expanded demand for R&D, design, marketing, trading, and financial services in Hong Kong so much that its economy has continued to grow despite the loss of production jobs.

The current major terminal operators see Hong Kong as their primary business base and they take a regional rather than a purely local perspective. Hong Kong International Limited (HIT) has moved further than the other operators in this "regional" regard. Through equity joint ventures, HIT is a leader in developing the ports in Yantian, Shanghai, and the Pearl River Delta.[1] Modern Terminals Limited (MTL) was also interested in Yantian but it has lost in its bid to HIT. It aims to link up with Shekou via its parent company, Wharf, and has been exploring opportunities to form joint ventures with the ports of Ningpo, Qingdao and Dalian. The other existing and future terminal operators like Sea-Land Orient Terminals (SLOT), COSCO, and some shareholders of the Asia Container Terminals (ACT), have been looking further north up the Chinese coast; they have focused primarily on small feeder ports. Some shareholders of ACT do not actively pursue the development of container ports in South China, because in their view the window of opportunity has been lost due to the fact that others have already moved in to fill the void.

If HIT and other Hong Kong firms had not participated in the development of Chinese ports, one might ask whether these ports would have remained undeveloped. In that case Hong Kong might maintain its unique position as the only modern port along the Chinese coast. The answer to this question is almost certainly not. The development of foreign trade and the expansion of cargo sources loudly demanded the development of modern ports in China. Had HIT and other terminal operators in the territory not been involved in developing China's ports, terminal operators from other parts of the world would likely have snapped up the opportunities. That is to say, given the changes in the economic landscape, it would be futile for Hong Kong, out of its own selfish motive, to try to block the development of regional ports. Instead, it would be more productive to be a partner of change and to identify win-win solutions.

Potential Major Competitors in the Region

Yantian and Shekou

Yantian, located on the eastern side of the Shenzhen Special Economic Zone and only 3 kilometres from the Sha Tou Jiao border crossing in northeast Hong Kong, has the highest potential among all South China ports to develop itself into a major container port. The port started operations in July 1994, and its throughput in 1995 was over 100,000 TEUs. The throughput for 1996 may exceed 300,000 TEUs. With two container berths (each with 350 metres of berth length and 14 metres of draft) on 50 hectares of land, the design annual capacity (existing facilities) is 500,000 TEUs.

In anticipation of fast expansion of its throughput, Yantian has planned to start construction at the end of 1996 on three new ship berths on fifty-two hectares of additional land, with a design capacity of 1.2 million TEUs. When the project is completed in 1999 the total design capacity of Yantian will reach 1.7 million TEUs. Many managers in Yantian are hired from Hong Kong and they basically use the same operations systems as the one used at HIT. Employing HIT's techniques, these managers may raise the actual capacity to 3 to 3.5 million TEUs.

The natural conditions of the port of Yantian are very favourable, and in the government's long-term plan, up to forty ship berths can be constructed. Thus, the speed of Yantian's development depends primarily on the pace of cargoes expansion. The volume of cargoes depends on the number of shipping lines that call the port and their frequency of call. Whether a port is going through a virtuous or vicious circle is most critical.To generate a virtuous process, the Yantian International Container Terminals (YICT) has connected itself to Hong Kong and coastal cities in northern Guangdong and Fujian Provinces by frequent feeder services to supplement its natural cargo base in Shenzhen and the adjacent areas.[2] It also plans to extend its feeder network to Hutchison's Pearl River Delta ports.Although many Chinese cities are linked to Yantian through a connecting line of the new Beijing-Kowloon Railroad (completed up to Shenzhen), it is not clear how many containers can be moved from the interior by rail.

YICT has launched promotion campaigns aimed at shippers and shipping lines and has been quite successful in attracting some major shipping lines. Two of the four shipping consortia, namely, Maersk/Sealand and the Global Alliance, as well as a number of major shipping lines, are already calling the port regularly. Momentum seems to be gathering, and other consortia and major shipping lines might follow suit. Judging from statistics available in late 1996 and early 1997, growth of throughput has been very rapid within a short period of time.

Like Yantian, Shekou is a port just a few kilometres away from the Shenzhen-Hong Kong border, but to the northwest. Its two container berths handled 90,000 TEUs in 1995 and are expected to handle 120,000 TEUs in 1996.[3] To promote itself to shippers, it signed a harbour agreement with the Grand Alliance in June 1996 for their ships to call at the port,[4] and it plans to promote barge services between it and the feeder ports of the Pearl River Delta. Shekou enjoys an advantage in river trade relative to Yantian, because Shekou is right at the mouth of the Pearl River. Partly offsetting this geographical advantage is that cargoes from Fujian and northern Guangdong are fed to Yantian with barges. Its most serious disadvantage, however, is its shallow waterway. Whether this disadvantage can be negated depends on whether investment will be made to dredge the Tonggu channel to the west of Lantau.

Advantages and Disadvantages

Relative to Hong Kong, both Yantian and Shekou enjoy two major advantages for cargoes originating in or bound for South China: (1) shorter trucking time due to not having to cross the Shenzhen-Hong Kong border and (2) lower tariffs. Thus, the factory-to-ship costs incurred by shippers are lower when using these ports than when using the port of Hong Kong. As reported in the Port Development Board (PDB) 's *Port Cargoes Forecasts 1995* (hereafter the "PDB Report", pp. 5.26-5.27)," carriers are very sensitive to pricing issues, and are aware of the through cost differences between the direct move from a PRC port and the move through Hong Kong.... Calling direct in the PRC gives them an opportunity to make more of a margin, while actually reducing the final product price to their customer."

Costs

According to the PDB Report (its Table 5.8), the terminal-handling charges (THCs) for a forty-foot-equivalent (FEU) container from Hong Kong and Yantian to the east coast of the U.S. are US$285 and US$192, respectively.[5] That is, the savings in THC per FEU is about US$100. The other major savings include US$50 from declaration at the Hong Kong border, US$50 in reduced import duties, and US$220 in trucking costs. Thus, the total savings amount to$420, not enough to offset the Yantian Arbitrary of US$450. Should the Arbitrary be abolished by 1997, as it is predicted that it will be, the net saving to the shippers would be US$420 per FEU.

According to YICT, the Arbitrary in August 1996 was only about US$250, and the savings in trucking costs was US$400, implying a net saving of US$350 per FEU. According to *Shippers Today* (September-October 1996, p. 33), the savings in trucking costs, THCs, and declaration/duties were US$380, US$103, and US$60, respectively. Thus, the total amount saved was $543, which, after a deduction of $300 as payment to the Port of Yantian, yields a net savings of $243 per FEU.

A disadvantage of Yantian is that much of the region's industrial output is in the west Pearl River Delta (PRD). "As Hong Kong is well connected with the PRD by river, and Yantian is not, the river trade represents a major advantage to Hong Kong in competition with Yantian", concludes the PDB Report (p. 5.67). This distinction is itself non-controversial, but, as pointed out earlier, Yantian plans to start feeder services to the Pearl River Delta to overcome its geographical disadvantage. Moreover, it also enjoys an advantage in that it receives cargoes from Fujian, northern Guangdong, and even east China by barges.

The PDB Report has also found that "it is usually simpler, cheaper, and faster for a container . . . to move to the hub port overland, if an adequate road system exists." (p. 5.23) An implication is that perhaps the advantage of having direct river-craft access to the Pearl River Delta might not be as important as are the savings in overland trucking costs enjoyed by Yantian and Shekou. After all, until now all containers that come to Hong Kong overland must go through Shenzhen, and so it would be cheaper and faster to go to the Shenzhen ports instead.

Thus, if shipping lines were to call Yantian as frequently as they call Hong Kong, then in terms of factory-to-port cost alone there would be every reason for shippers to shift cargoes originating in the Pearl River Delta from the port of Hong Kong to the ports of Yantian and Shekou.[6]

Customs and Regulation

As much as Yantian and other ports in South China enjoy the twin advantages of lower tariffs and shorter trucking time, they are at a serious disadvantage in terms of customs clearance, government regulations, and the quality of related services. The inefficiency of customs clearance and other regulatory procedures is well known, even though the various ports try to improve efficiency by providing a one-stop service for customs clearance. But as the PDB Report (p. 5.34) puts it, "it still represents a critical area of concern and threatens on time delivery for China origin cargoes. While most survey respondents accepted that the customs environment would restructure and improve to an acceptable level, most believed that this would take at least 5 to 10 years." Even optimistic terminal operators feel that it will take no less than five years before these ports attain international standard in customs clearance and regulations. However, an industry source has revealed that significant progress has been achieved recently and there is reason now to be less pessimistic.

One might argue that due to Yantian and Shekou's location in Shenzhen, strict parity in customs and regulatory efficiency is not necessary to make these ports competitive vis-a-vis Hong Kong in terms of the total time required to reach their respective terminals. Containers coming to Kwai Chung from China overland have to clear customs twice, once at the border and once at the port of Hong Kong. Thus, so long as customs clearing at the Chinese ports is as quick as is that at both the border and the port of Hong Kong plus additional transport time to the latter, no additional delay will have resulted from using Chinese ports.

Port operators in China understand the importance of things other than physical structure and equipment to making their ports attractive to shipping lines and shippers. For example, according to a report, Shekou "would concentrate on improving customs and other soft infrastructure procedures including concentrated customs declaration, electronic data interchange, development, container inspection ratios, commissioning of a one-stop documentation centre and transshipment clearance of import and export cargoes."[7] Based on their land and labour cost advantages, some of the Chinese ports might become fierce compeitors of Hong Kong if they can achieve high productivity, offer high-quality services, and meet world standards in customs and government regulations.

"Lock-in" Effects and Economy of Agglomeration

As pointed out at the beginning of this chapter, established ports have a significant advantage over new ports by virtue of the "economy of agglomeration". In the beginning, shippers are reluctant to use South China ports instead of Hong Kong's port because of the importance they attach to speedy and reliable delivery. The PDB Report (p. 5.35) has found that "consolidators operating in Hong Kong and Guangdong have experienced client hesitation to shifting to FOB (PRC) terms of sale and consolidation in Guangdong even after demonstrating that the services . . . are comparable and cheaper than those available in Hong Kong."

The above finding shows that business confidence takes time to develop. However, given the increasing number of international shipping lines calling South China ports, there might not be as much inertia as there was once thought to be. Indeed, some believe that Yantian has reached the critical point at which a virtuous circle might set in soon.

Due to the high costs of calling every port, a single shipping line typically faces the choice of calling some ports versus others. Through the pooling of ships, a shipping alliance can call many ports at a lower total cost than that which a single shipping line will incur for covering the same ports. Therefore, the alliance will typically find it economical to call more ports than a single shipping line will. As a result, the formation of alliances has weakened the economy of agglomeration enjoyed by established hubs. Equivalently, new and smaller ports with acceptable service quality and a cost advantage will face less serious hurdles in the start-up process; and they may be able to exploit their cost advantage to a degree that would not be possible without these consortia. Once enough cargo volume is built up, the range of supporting services at the new ports can be expected to expand.

Net Effect

The net effect of the advantages and disadvantages of the South China ports vis-a-vis Hong Kong can be seen through the choice of shipping lines and shippers. An increasing number of shipping lines are making calls to Yantian and other South China ports. According to the PDB Report (p. 5.29) "there is an increasing momentum towards overseas buyers specifying FOB China, with many leading forwarders and US stores already using or considering a shift to these terms."[8] Furthermore, "as Hong Kong's advantages as the only international business centre in the region become less dominant, cargo diversion from Hong Kong to South China becomes more likely." (p. 5.31)

Despite this development, experts in the industry believe Hong Kong will remain the major hub for years to come, while Shekou to the west and Yantian to the east will serve mainly as subsidiary ports. Their belief reflects the recognition of the formidable natural and man-made advantages enjoyed by the port of Hong Kong. To be sure, any statement about the erosion of Hong Kong's dominance or the regional ports' increasing competitiveness should not be interpreted as an absolute decline in cargoes handled by Hong Kong. No one is predicting the demise of Hong Kong as a hub port, but many believe that it will handle a diminishing share of the region's total port cargoes.

Gaolan

Gaolan in Zhuhai lies to the southwest of Macau. Unlike Shekou and Yantian, it is nowhere near Hong Kong. While its natural conditions are more favourable than are Shekou's, they are less favourable as compared with Yantian's. Thus, Gaolan's chance of developing into a major regional container port is not as great as is

Yantian's. Some think Gaolan might remain largely a feeder port in the foreseeable future.

Kaohsiung

In the last several years, Kaohsiung's container port has become the third-busiest in the world, far behind Hong Kong and Singapore but slightly ahead of Rotterdam.[9] It has built many ship berths and has an ambitious plan for expansion. It now boats a total of twenty-two berths, with draft ranging from 10.5 metres at Container Terminal No. 1 to 15 metres at Terminal No. 5. The berths' lengths vary from below 200 metres to close to 440 metres.[10] In addition, five additional berths have been planned or are under construction. Even with an average of 300,000 TEUs per berth, the total capacity would be 6.6 million TEUs; at 400,000 TEUs per berth the capacity would be 8.8 million TEUs, well above the current throughput of 5.05 million TEUs. Kaohsiung is an example of a container port in which capacity is not a binding constraint on throughput growth. In any event, the Kaohsiung International Port Overall Development Project aims to expand its total capacity much higher, to 23 million TEUs by 2030.

Kaohsiung's major advantages vis-a-vis Hong Kong are twofold: it enjoys a central geographical location in East Asia and it retains lower costs.[11] Its two major disadvantages are its cumbersome regulatory procedures and its lack of direct trade with China.[12] A number of reform measures have been introduced to streamline and rationalize regulations, and "offshore shipping centres" have been proposed to deal with transshipment involving China in such a way that ships can run between Taiwan and ports of Mainland China legally but "without involving customs clearance or entry into Taiwan Area." The PDB Report expects that 90% of cargoes routed through Hong Kong will soon be lost if direct trade links between Taiwan and China are established.

Shanghai and Ningpo

The Shanghai Container Terminals Limited (SCT), established in

1993, is a joint venture between Shanghai Container Terminals Comprehensive Development Co. and Hutchison Ports Shanghai Limited. It operates three container terminals converted from traditional general purpose terminals along the northern stretch of the Huangpu River, namely, Baoshan Terminal, Zhanghuabin Terminal, and Jungonglu Terminal. Altogether there are now a total of ten berths with a total quay length of 2281 metres; the draft varies from 10.5 to 12.5 metres. The terminals can handle container ships with a capacity of up to 3800 TEUs but will not be able to handle larger container ships without significant investment in dredging and maintenance.

The total design capacity of the existing berths at SCT is 1.7 million TEUs. If appropriate equipment and operation systems are used, then the total capacity could be much larger, say 3 to 4 million TEUs. SCT started operating in August 1993. Its throughput in 1994 and 1995 was 1.13 million TEUs and 1.29 million TEUs, respectively, representing 25.5% and 14% annual growth from the respective previous years. Thus, given the potential operating capacity, it will take some years before Shanghai could approach its capacity limit.

As a container port, Shanghai's advantages and disadvantages are well known. Located at a vital juncture of the booming Yangtze region, it enjoys access to a huge hinterland for cargoes. However, the shallowness of its port will confine it to smaller container ships, though handling the largest container ships in the near future is not necessarily a pressing need. Its tariffs are low not only relative to Hong Kong, but also relative to Yantian.[13] Like other Chinese ports striving to turn their cost advantage into a real competitive advantage, SCT is working hard to improve its service quality, including turn-around time,[14] flexibility in the lead time for accepting shipments, customs clearance, and the availability and quality of supporting services.

Most of SCT's cargoes represent direct shipment. Like Yantian, transshipment accounts for only a small percentage, but its relative importance is expected to rise. With its catchment area being the

Yangtze region which is far away from Hong Kong Shanghai is not a direct competitor of Hong Kong in the area of container handling.

In contrast to Shanghai, Ningpo has the favourable natural conditions as a deep-water port. With the help of the tide and dredging, its approach channel can accommodate ships of 300,000 tonnes. The section at Beilun has especially favourable natural conditions: there is deep water over a large and well-sheltered area of the sea. With a draft of 17.6 to 21 metres along theapproach channel and 18 metres by the terminals, Beilun has the potential to handle the largest container ships.

At the container terminal, there are currently four berths with a total of 900 metres of quay and more than 14 metres of draft. A joint venture with COSCO has converted 320 metres of the quay into one berth that can handle third- and fourth-generation ships (1000-2000 TEUs, and 2000-4000 TEUs, respectively). It is looking for a second partner to convert the remaining 580 metres of quay into two additional berths. Discussion with Wharf in Hong Kong began in 1992, but so far no agreement has been reached.

In addition to converting the existing berths, Beilun plans to build four new berths with a total length of 1238 metres and a design capacity of 2 million TEUs during the Ninth Five-Year National Plan. Together with the converted berths, total capacity will reach 3 million TEUs.

In contrast to Shanghai, the binding constraint for Ningpo is its cargo source. Its catchment area is relatively small due to underdeveloped transport links to China's economic hinterland. The container throughput in 1995 was about 160,000 TEUs, and that in 1996 is expected to reach 200,000 TEUs. A key strategy for Ningpo's expansion of its cargo base is to link up with Shanghai, with a strong urging from the central government.[15]

In the long run, Ningpo's natural advantages can remedy Shanghai's natural limitation, i.e., the shallowness of the Huangpu River and the nearby sea. However, Shanghai will not reach its capacity limit for quite some time, and there are plans to build new container terminals in the Pudong New Area. Thus, a joint venture between the two regional governments brokered by Beijing might not bode well when it is not yet imperative for Shanghai to handle the largest of container ships.

Challenges and Opportunities

To sum up, Yantian has the potential to pose a serious challenge to Hong Kong. However, in the many years until Yantian's potential is actually realized, Hong Kong's container port will continue to enjoy its natural, man-made, and systemic advantage. It will probably take quite a long while for Yantian to develop its throughput to a level that rivals that of Hong Kong; so the territory may be able to retain its dominant position for another decade or longer. On a positive note, there is an opportunity to integrate Hong Kong, Yantian, and Shekou into a mutually supportive port system.

To the territory's container handling industry, the threat of Kaohsiung is serious, but the threat is mainly in terms of transshipment into and out of Chinese ports. Once direct trade between Taiwan and China is permitted, Kaohsiung will present a more serious challenge to Hong Kong's transshipment business than do Shanghai and Ningpo, because Kaohsiung is not only closer to Hong Kong but is also more developed than ports in China as a world-class container port.

In summary, we have reviewed port development in the region and have discussed the threats and opportunities the regional ports present to Hong Kong. In the following two chapters, we turn to the container handling capacities of the territory's existing and planned facilities and to the demand for the territory's container handling services during the next ten years.

Notes

1. Currently there are five Pearl River Delta ports managed by Hutchison Delta Ports, namely, Jiuzhou and Gaolan in Zhuhai, San Shan in Nanhai, Shantou, and Jiangmen. With the exception of Gaolan, the others would remain as feeder ports. Three more ports might be added in the future.
2. Yantian is connected to major cities of the Pearl River Delta by a road network. A highway connects it to Huizhou and the centre of Shenzhen. It is also connected to Guangzhou and the surrounding area by the Guangdong Superhighway. Additional highways are being planned to broaden Yantian's catchment area and to bypass downtown Shenzhen.
3. *South China Morning Post*, 10 April 1996.
4. *South China Morning Post*, 26 June 1996.
5. According to experts in the industry, these THCs are close to the tariffs charged at the two ports.
6. If the shallowness of Shekou's approach channel is remedied.
7. The report "Shekou threat to Hong Kong transshipments" appeared in the *South China Morning Post*, 10 April 1996.
8. A footnote on the same page gives an impressive list of major retailers that have adopted or are considering the adoption of FOB China. In addition to cost differences, two other developments have been given as reasons for direct calls of Chinese ports. First, China has been granting shipping lines branch office status, thus breaking the monopoly of state-owned freight forwarders. Second, a number of Hong Kong freight forwarders are allowed to do business in China. (*South China Morning Post*, 14 March 1996, "Direct routes signal challenge to Hong Kong port volumes".)
9. Kaohsiung handled 4.6 million TEUs, 4.9 million TEUs, and 5.05 million TEUs in 1993, 1994, and 1995, respectively. The corresponding figures for Rotterdam were 4.3 million TEUs, 4.5 million TEUs, and 4.78 million TEUs.
10. Kaohsiung Harbor Bureau (1996).
11. Ironically, some shipping lines regarded "relatively high and rising costs" as a disadvantage of Kaohsiung (Council for Economic Planning and Development (1994, 2/19)). Some shipping companies also claimed that it suffered from unpredictable costs because the "port authorities unilaterally raised fees substantially." (2/32)
12. There are many inconvenient procedures and restrictions on using facilities. For instance, leased terminals are forbidden from leasing surplus capacity to other shipping companies, and lessees are assessed

extra fees for sharing dock facilities. In addition, long lead times are required by the authorities for processing documents.

13. John Meredith, managing director of HIT, put the tariff at US$66 per TEU, according to the *Asian Wall Street Journal*, 26 March 1996. It was revealed that the saving per TEU to Los Angeles from Shanghai as opposed to from Hong Kong was about US$300 to US$400 (*South China Morning Post*, 14 March 1996).

14. In 1995 the average time of handling a ship (for ships up to 3800 TEUs) was 13 hours, and an effort is being made to reduce it further.

15. According to officials at Beilun, in a speech about establishing Shanghai as a international transport center in January 1996, Premier Li Peng asked Shanghai and Ningpo to form a joint venture to handle shipping business.

CHAPTER 6

Capacity of Existing and Planned Facilities

Capacity of Kwai Chung and Tsing Yi Terminals

What is the total handling capacity of the existing terminals in Kwai Chung and Terminal 9 to be built on Tsing Yi? Opinions on this issue differ substantially.

The Port Development Board (PDB) has reported that, according to "a review finalized in May 1995 . . . , the annual throughput capacity of the existing container terminals (CT1 to 8) was 9.15 million TEUs."[1] In addition the PDB predicts that the capacity can increase by another 17% to a ceiling of 10.7 million TEUs by 1998.

Kwai Chung terminal operators' estimates of their own current capacities seem to be consistent with PDB figures. Hongkong International Terminals (HIT) boasts a total annual handling capacity of 4.5 million TEUs, and COSCO-HIT has a capacity of 1 million TEUs.[2] Modern Terminals Limited (MTL)'s capacity is 2.5 million TEUs,[3] and Sea-Land Orient Terminals (SLOT)'s capacity on Terminal 3 alone is set at about 1 million TEUs. Hence, by their own account, the total capacity of the existing terminals is about 9 million TEUs, similar to that estimated by the PDB and a local consulting firm for 1995.[4]

However, the PDB's estimate of the terminals' future capacity is at odds with what operators and some industry analysts believe is possible. For instance, while the PDB put the figure at 10.7 million TEUs for 1998, James Capel Asia estimated that the capacity would reach 11.55 million TEUs in the same year, and some operators believed that the capacity in 1996 was already at 11 million TEUs and that with additional backup land it could reach 12 million TEUs by 1998.

The capacity estimate of 12 million TEUs is about 12% higher than that provided by PDB. Is the higher estimate reasonable? Since there are a total of eighteen ship berths on Terminals 1 to 8, a total capacity of 12 million TEUs implies an average capacity of about 666,700 TEUs per year per berth. That is substantially higher than the actual average throughput per berth in 1995; but it is definitely feasible, because the single berth on Terminal 3 handled 880,000 TEUs in 1995 and more than one million TEUs in 1996.[5] Although SLOT is different from other operators in that it is to a large extent vertically integrated with the shipping line Sea-Land, so that it is less constrained by demand compared with other terminal operators, the technological potential for handling 666,700 TEUs per year per berth has clearly been demonstrated, provided that adequate backup land (say, fifteen hectares per berth) is available to operators.

Now that the agreement to swap berths has been finalized, the four berths on the new terminal might begin operation in the second half of 2000. How much additional handling capacity will they bring on stream? The design capacity figure used by the PDB is 550,000 TEUs per berth, but it is clearly on the low side. Each of the four new berths has 320 metres of quay front and fifteen hectares of land, and an alternative assumption of 700,000 TEUs per berth with a total capacity of 2.8 million TEUs does not seem off the mark, especially because the two barge berths at the north and south ends of the terminal are no different from regular ship berths (except for the fact that they have much less backup land) and can be used as such. Indeed, some industry experts believe that if the barge berths are used as ship berths, then the total capacity of Terminal 9 may reach 3.5 million TEUs by 2004.

In summary, we assume that the total capacity of Terminals 1 to 8 will reach 12 million TEUs by 1998 and will stabilize at that level afterwards.[6] In addition, we assume that the new berths on

Terminal 9 will take about two years to reach their maximum capacity of 2.8 million TEUs. From 2004 to 2006 it may reach 3.5 million TEUs. Based on these assumptions, the total handling capacity of the terminals in Kwai Chung and Tsing Yi over the next ten years would be shown in Table 6.1.

The construction of Terminal 9 is expected to take twenty-seven months. If construction begins in the last quarter of 1997 the first berth would become available for use by March 2000, and all ship berths could be ready by the end of year 2002. They would contribute to the total capacity in 2000 by 1.5 million TEUs.

Capacity of Mid-Stream Operations

Just as the container terminals, the total capacity of mid-stream operations depends on the number of sites and support facilities made available to such operations. Recall from Chapter 2 that mid-stream operations handled almost 3 million TEUs in 1995. Besides the existing quay front and a permanent site on Stonecutters Island to be ready for use in 1997, other potential areas include Tseung Kwan O, Northshore Lantau, Lung Kwu Tan in Tuen Mun, and east of Terminals 10 and 12 planned for Lantau.[7] Informed industry sources put the capacity of mid-stream operations for 1996 at 3.5 million and think that the total capacity can grow to over 6 million in a decade if land and working areas are added.[8]

Assuming that the capacity of mid-stream operations increases by 500,000 TEUs every two years, the total capacity of Terminals 1 to 9 and mid-stream operations is given in Table 6.2.

The real question is whether mid-stream operations are best regarded as (a) providing a safety valve when existing container terminals face capacity constraint as new terminals are being built, or (b) permanent facilities to be expanded along with terminals. If the view (a) is taken, say because dedicated terminals are socially more efficient than mid-stream operations are, then mid-stream facilities would not continue to expand beyond some given level, especially when the government has difficulty finding new mid-stream sites to replace or supplement old sites.[9] Looking at the

Table 6.1
Capacity of Terminals 1–9

	Million TEUs in					
Terminals	1996	1998	2000	2002	2004	2006
1–8	10–10.5	12	12	12	12	12
9			1.5	2.8	2.8	2.8
1–9	10–10.5	12	13.5	14.8	14.8	14.8

Table 6.2
Capacity of Terminals 1–9 and Mid-Stream Operations

	Million TEUs in					
	1996	1998	2000	2002	2004	2006
CT 1–9	10–10.5	12	13.5	14.8	14.8–15.5	14.8–15.5
Mid-stream	3.5	4	4.5	5	5.5	6
CT 1–9 & Mid-stream	13.5–14	16	18	19.8	20.3–21	20.8–21.5

Table 6.3
Capacity of Terminals 1–9 and Mid-Stream Operations

	Million TEUs in					
	1996	1998	2000	2002	2004	2006
CT 1–9	10–10.5	12	13.5	14.8	14.8–15.5	14.8–15.5
Mid-stream	3.5	4	4	4	4	4
CT 1–9 & Mid-stream	13.5–14	16	17.5	18.8	18.8–19.5	18.8–19.5

situation from that point of view, the large size of mid-stream throughput signals a need to build new container terminals.

If, for instance, the maximum is 4 million TEUs, then the total capacity of Terminals 1 to 9 and mid-stream operations would be reduced accordingly as shown in Table 6.3.

Capacity of River-Trade Terminals

In terms of the origin and destination of cargoes, river trade is different from ocean cargoes handled at container ship terminals and in mid-stream operations. Facilities used for handling river-trade cargoes are complementary to those used for mid-stream operations and main terminals. Thus, neither the demand nor the handling capacity of river-trade should be mixed with those for ocean cargoes. According to a recent report, the Tuen Mun river-trade terminal is expected to have sufficient capacity to handle one million TEUs by 1999.[10]

Capacity of South China Ports and Their Implications for Hong Kong

At present the existing throughput of the South China ports is very small. For instance, the throughput for Shekou was 90,000 TEUs in 1995 and is expected to rise to 120,000 TEUs in 1996. Yantian handled over 100,000 TEUs in 1995 and is expected to reach 350,000 TEUs in 1996. The total projected throughput of these two nearby ports in 1996 amounts to less than 3.5% of Hong Kong's 1995 throughput.

As we recall from Chapter 5, the capacity of these two ports is much larger. Shekou has two container berths with a total design capacity of 500,000 TEUs, and two new berths are expected to be completed by the end of 1996, bringing its total capacity up to one million TEUs. Yantian has two container berths with a total design capacity of 500,000 TEUs, and three new berths are expected to be completed by 1999, bringing its total design capacity up to 1.7 million TEUs.

What are the implications of Shekou and Yantian's design capacity for Hong Kong's port facilities requirements? The answer is that unless they can attract enough cargoes to feed their ports, an increase in their physical capacities does not necessarily imply a one-to-one reduction in Hong Kong's required facilities.

Even though Yantian can physically reach a capacity of 8

million TEUs in ten years, to what extent this potential is realized depends critically on the dynamics of shippers' demands. To appreciate the time it might take Yantian to develop into a major container port, it would be helpful to use other major ports in the world as points of reference. Kaohsiung, the world's third-busiest container port, with excellent natural conditions, handled slightly more than 5 million TEUs in 1995 after many years of development. Rotterdam, the fourth-busiest container port in the world and the largest in Europe, handled 4.8 million TEUs in the same year.

The likelihood that the South China ports might take a long time to develop before causing serious cargo diversion implies that, in the next ten years or so, Hong Kong's throughput will continue to increase, albeit at slower rates than it has in the past. It would not be very meaningful to subtract the total maximum technical capacity of the South China ports from total container throughput of Hong Kong and South China to arrive at a residual that is interpreted as Hong Kong's required capacity. The flaw in this type of analysis lies in its ignoring the dynamics of cargo development and in equating the demand for South China port facilities with their maximum technical capacity.

Equally wrong would be to assume that whatever capacity created in Hong Kong will be fully utilized, so that only the demand for container handling not met by Hong Kong would be available to South China ports as an overflow. The truth of the matter is that each port's capacity only sets an upper limit on its throughput, but the actual distribution of cargoes among them depends on how well they meet the service-demand of shippers and shipping lines.

Having rejected the approach of subtracting the maximum technical capacity of a subset of ports from the region's total throughput to arrive at the demand for capacity at other ports, we shall in the next chapter make assumptions directly about demand for Hong Kong's container handling services.

Notes

1. PDB, *Annual Report for the period up to September 1995*, p. 8.
2. *HIT Facts Sheet.*
3. The capacity figure was provided in a report in the *South China Morning Post*, 21 May 1996.
4. According to a report in *South China Morning Post*, 11 January 1996, James Capel Asia put the capacity estimates for HIT, COSCO-HIT, MTL, and SLOT at 4.4 million, 1.3 million, 2.65 million, and 0.8 million TEUs, respectively. The figures added up to 9.15 million TEUs, the same as the PDB estimate of total capacity. But some people in the industry think that the capacity in 1995 was already in excess of 11 million TEUs.
5. According to the *South China Morning Post*, 11 March 1996, SLOT only had a 70% utilization rate then. MTL expects its five berths to have a handling capacity of about 3 million TEUs in 1997 and 3.3 million TEUs in 1998.
6. Some in the industry believe that their total capacity from 2002 to 2006 can reach 12.5 million TEUs.
7. *Port Development Strategy Second Review*, Executive Summary, October 1995.
8. Some in the industry even believe that the capacity of mid-stream in 1995 could be as high as 5 million TEUs.
9. According to a PDB study, Economic Appraisal of Mid-Stream Operations (April 1991, p. 13), "the capital costs of the container terminals are only marginally higher than the equivalent mid-stream buoy systems, in the range of 2% to 4%," but "the container terminals' operating costs are on average approximately 60% of those of the equivalent mid-stream system." It thus follows that from a social point of view, terminals are more cost effective than are mid-stream systems at reasonable discount rates. Notice that in the study "costs" are defined as resource costs, regardless of whether they are incurred by the government, operators, or users of the latter's services. As a result, the premium paid by terminal operators for the right to develop terminals is **not** included, so the costs do not reflect the private costs borne by operators.
10. *Ming Pao*, 17 October 1996.

CHAPTER 7

Future Demand for Container Handling Services

The purpose of this chapter is to critically assess the forecasts of container throughput as provided in the Port Development Board (PDB) Report, *Hong Kong Port Cargo Forecasts 1995*, in light of (a) actual throughput figures that were not available when the report was prepared, (b) updated information about the Gross Domestic Product (GDP) or Gross National Product (GNP) growth rates of Hong Kong's major trading partners, and (c) alternative assumptions about the growth of container throughput, including those about the Port Rail Link.

Without details of the parameters used in PDB's forecasting model, we are not in a position to comment on the computational aspects of how these figures were derived. We take the PDB figures as given, but examine how they would be affected by the new information, and present alternative estimates for ocean container throughput only up to 2006 instead of through 2016, the last year of the PDB forecasts.

There are two reasons for not going beyond 2006. First, there is greater uncertainty the farther we gaze into the future. It would be difficult enough to make sensible assumptions about the growth of cargo volume more than ten years from now. To do so about diversion of cargo from Hong Kong to regional ports would be even more difficult. The latter difficulty is due to uncertainty about what kind of competitors Yantian, Kaohsiung, Shanghai and Ningpo

Table 7.1
Forecasts of Ocean Container Throughput (1000 TEUs): 1996–2006

	1996	1997	1998	1999	2000	2001	2002	2003	2004	2005	2006
PDB's 1995 forecasts	12,650 (11.8%)	14,156 (11.8%)	15,826 (11.8%)	17,060 (7.8%)	18,391 (7.8%)	19,824 (7.8%)	20,776 (4.8%)	21,773 (4.8%)	22,818 (4.8%)	23,914 (4.8%)	25,005 (4.8%)
Apply PDB's original growth rates to the 1996's actual throughput	11,726	13,110 (11.8%)	14,656 (11.8%)	15,800 (7.8%)	17,032 (7.8%)	18,361 (7.8%)	19,242 (4.8%)	20,166 (4.8%)	21,134 (4.8%)	22,148 (4.8%)	23,211 (4.8%)
Use 1996's actual throughput and alternative throughput growth rates (in parentheses)	11,726	12,781 (9%)	13,932 (9%)	15,046 (8%)	16,250 (8%)	17,387 (7%)	18,604 (7%)	19,720 (6%)	20,904 (6%)	22,158 (6%)	23,488 (6%)
Drop bullish cargo figures produced by KCRC from original PDB forecast, and adjust growth rates after 1996 based on new GDP/trade growth rates not available at the time the PDB forecasts were made	11,726	13,075	14,566	15,642	16,799	17,651	18,470	19,331	20,238	21,195	22,152
Add back modified KCRC estimates (delayed by two years and adjusted according to these fractions: 50% for 2003, 70% for 2004, 90% for 2005 and 2006)	11,726	13,075	14,566	15,642	16,799	17,651	18,470	19,545	20,605	21,758	22,805
Apply PDB's March 1997 forecastedl growth rates to the 1996's actual throughput	11,726	12,723 (8.5%)	13,804 (8.5%)	14,977 (8.5%)	16,251 (8.5%)	17,632 (8.5%)	18,778 (6.5%)	19,999 (6.5%)	21,298 (6.5%)	22,683 (6.5%)	24,157 (6.5%)

will develop into a decade or two hence. Drastically different scenarios may emerge over that kind of time horizon. It is also difficult to predict whether container handling will remain in Hong Kong's comparative advantage vis-a-vis other kinds of economic activities such as finance and high-tech industries. Success in other activities without corresponding productivity increases in cargo handling will squeeze out the latter.

The second reason for not venturing to provide estimates extending beyond 2006 is that a ten-year forecast is sufficient for addressing questions about the necessity of and timing for Terminals 10 and 11. At this point, there is no need to make irreversible decisions about Terminals 12 and 13.[1] Decisions about these terminals can be made in due course with the arrival of data in the next five years. We do recognise that because of economies of scale in developing the Lantau Port Peninsula, the decision to build Terminals 10 and 11 cannot be entirely separated from Terminals 12 and 13.

Given that the first dedicated river-trade terminal will not be completed until 1999, there is no basis upon which to assess the performance and adequacy of river-trade terminals. Thus, in this chapter, we focus on ocean container throughput only.

Forecasts of Ocean Container Throughput (1997–2006)

In Table 7.1, six sets of ocean container throughput figures are presented. The first set, in the top row, reproduces the 1995 forecasts of the PDB Report (Table D1 in *Appendix XIII*). Since the report provides forecasts for only 1998, 2001 and 2006, we obtain the forecast of the other years within any given period by assuming that the annual growth rate within the period is equal to the "average annual growth rate" for that period is as provided in Table D1. That is to say, our figures would exhibit rougher changes than PDB's which were based on sliding growth rates within any given period.

The second row provides the minimal modification of the PDB forecasts. The actual throughput in 1996 was equal to 11.726 million TEUs, 8.686 million TEUs of which were handled at the terminals. In all other years, we use the same annual growth rates as in the first row. In other words, we assume that the drastically slower growth in container throughput in 1996 (namely, 5.2% growth at the terminals and 3.8% growth in mid-stream operations) was an aberration rather than a signal of a downward growth trend.

There are reasons to suspect that the growth trends have changed, partly due to a slow-down of GDP and trade in the world, and due in particular to a realization that the Asian dynamic economies have shown signs of "maturing". Many in the industry feel that future growth rates in container throughput are likely to be single-digit rather than double-digit. The figures in the third row of the table are based on the 1996 figure from the second row, with the assumption that ocean container throughput would grow at an annual rate of 9% during 1997–1998, 8% during 1999–2000, 7% during 2001-2002, and 6% during the rest of the forecasting period. Some in the industry believe that 7% is a likely growth rate from 1996 onwards, in which case the throughput figures would be even lower.

The figures in the fourth row were obtained in a completely different manner. First, the very bullish forecasts on the number of containers brought from the interior of China to Hong Kong by rail as provided by the Kowloon-Canton Railway Corporation (KCRC) for 2001 to 2006 were dropped. The estimates provided by the KCRC for 2001, the year when the Port Rail Link was assumed to begin service, were 430,000 TEUs and 930,000 TEUs for 2006 (Table 5.15, *Port Cargo Forecasts 1995*, p. 5.71). Given that in 1995 the containers brought to Hong Kong from China by rail were less than 6,000 TEUs, largely because of problems inherent in the Chinese railway system with regard to the transport of containers, these estimates are very bullish and obviously unrealistic according to some critics. More fundamentally, it is not clear whether there will be a Port Rail Link, as was previously contemplated. But, in any event, it might be unrealistic to expect that a new railway will be ready to bring containers to the port of Hong Kong in 2001 as previously planned. The KCRC figures for between 2001 and 2006 were obtained by using simple interpolation and were then subtracted from the PDB figures in the first row. Second, based on the estimated ocean throughput for 1996 (11.707 million TEUs) and the original estimate used by the PDB (12,650), the forecasts from 1997 to 2006 were proportionately scaled down after the KCRC figures had been subtracted from the original PDB forecasts.

Third, the PDB annual growth rates implicit in the above figures were adjusted to reflect changes in GDP and trade growth estimates of Hong Kong's trading partners. Some estimates used by the PDB were explicitly stated in PDB's Table 5.22 and Appendix V. The estimated growth rates for most of Hong Kong's trading partners, however, were not revealed. In these cases, we used the differences between the forecasts of their GDP growth as given in the International Monetary Fund (IMF) report (May 1996) and/or the Organization of Economic Co-operation and Development (OECD) report (June 1996) and those that were available one year ago, when the PDB Report was prepared. Economists and growth watchers around the world are now reassessing the less-rosy growth prospects of the dynamic Asian economies,[2] but it will take some time before any significant downward adjustment of these economies' growth rate are reflected in new IMF and OECD forecasts.

In any event, we then translated the changes in GDP growth into changes in trade growth by using a factor of 1.2 (i.e., an increase in GDP by one additional percentage point will lead to an additional 1.2% growth in trade).[3] The adjustments to the trading partners' trade growth were then weighed by their respective shares of Hong Kong's outward ocean cargoes in 1995. The result of all these calculations indicated that the 1997 throughput growth rate should be adjusted downward by 0.57%, that the rate from 1998 to 2001 should be adjusted downward by 0.4%, and that the rate from 2002 to 2006 should be adjusted upward by 0.23. The resultant throughput figures are presented in the fourth row of Table 7.1.

In the fifth row, we attempted to add back the container cargoes that might be brought to Hong Kong by rail. For example, we supposed that a new rail system devoted to container cargoes would become available in 2003, but the amount of cargo might be substantially smaller than the KCRC figures indicate. The number of containers brought by rail would depend not only on China's railway system but also on what fraction will go to Yantian, a less-costly port that might have gained momentum by then. Thus, besides pushing the KCRC estimates back by two years, we assumed that only 50% of the estimate would be realized during the first year of operation in 2003, 70% in the second year (2004) and 90% in the third and fourth years (2005 and 2006). In the fifth row of Table 7.1, the estimates have included these containers.

As this book went to print (March 1997), the PDB adjusted its forecasted growth rates downward in light of the lower than expected growth in 1996. From 1997 to 2001 the new growth rate was 8.5% per annum; from 2002 to 2007 the new growth rate was 6.5% per annum. Assuming that these growth rates also apply to mid-stream throughput, we generate the last set of estimates in Table 7.1.

What is remarkable is that the last four rows of the table are in substantial agreement. The figures for 2002 to 2006 in the third and fifth rows were derived using very different assumptions, but they turned out to be quite close to one another. The differences between the third and last rows were directly the result of different assumptions about growth rates. In the third row the annual growth rate declined gradually from 9% in 1997 to 6% by 2003, whereas in the last row the annual growth rate declined from 8.5% in 1997–2001 to 6.5% in 2002–2006.

One can try other assumptions to generate other forecasts. As in any forecasting exercise, the forecasts are only as good as the assumptions behind them. We believe that the figures in the last four rows are adequate for illustrating the impact of key factors on the timing of new terminals beyond Terminal 9.

Comparison with PDB Forecasts

The forecasts contained in the last four rows of Table 7.1 are clearly lower than the PDB's original forecasts (first row) and updated forecasts (second row) for the simple reason that the PDB forecast for 1996 was too optimistic in light of actual containers handled. The PDB forecast for 1997 could be overestimated by 1 million TEUs and up to 1.5 to 2.8 million TEUs for 2006.

It is interesting that the third to fifth rows turn out to be quite close to what were termed "low growth scenario" forecasts as contained in Table 6.11 of the PDB Report. However, it should be pointed out that the figures in the fourth and fifth rows of Table 7.1 do not really represent a pessimistic scenario, because we have not incorporated all the negative developments that could occur.

Some of the assumptions adopted by the PDB in arriving at its "low-growth scenario" are of questionable validity. For example, it implicitly adopted an overflow model by assuming that if the growth of the cargo base was lower, then a smaller fraction of that cargo base would spill over to South China ports. However, a diversion of cargo could be due to increased competitiveness of the South China ports over and above what is currently expected. Thus, the possibility of a double-hazard that these ports might snatch a larger share of a smaller pie cannot be ruled out.

Another example illustrating the questionable validity of PDB assumptions arises from the assumption that the number of Chinese containers brought to Hong Kong by a new railway will be equal to 75% of the figures provided by the KCRC. But 75% of the KCRC figures may still be too optimistic.

Finally, in the PDB's low-growth scenario, container throughput continued to maintain positive growth up to 2016. However, given all the uncertainty about the competitiveness of regional ports such as Yantian, Kaohsiung, Ningpo, and Meizhou (in Fujian), and given changes in Hong Kong's own comparative advantage in container handling vis-a-vis other economic activities, the possibility that there will be a decline in twenty years' time cannot be ruled out. Thus, it is not clear how much importance one should attach to the forecast of 30 million TEUs of ocean containers in 2016.

The forecasts generated in the PDB's earlier exercises turned out to have underestimated actual demand. That was in part a consequence of the bottleneck that occurred in China's port facilities when foreign trade underwent rapid growth.The PDB's 1995 forecasts attempted to avoid its past mistakes by adopting more optimistic growth assumptions. Given the rapid development of ports in South China and China's policy of adopting a lower growth target to keep inflation down, one may wonder whether PDB's 1995 forecasts were overestimates. Even if the slow growth in 1996 was an aberration, one may still question whether the growth in the next ten years will be as robust as is implied by PDB's forecasts. If we compare the PDB's 1993 forecasts (its Table 6.7, p. 65) with the forecasts presented in the third to fifth rows of our Table 7.1, we see that the PDB's earlier forecast for 1996 was very much on target, though that for 2006 was slightly below our figures. Moreover, the PDB's forecast for 2001 was very close to the forecasts given in the third to fifth rows of Table 7.1. The figures in the last row reflect PDB's effort to correct for the 1995 overestimates.

In addition to the total volume of ocean containers, the throughput composition could also play a part in determining what types of container handling facilities are appropriate. Some people in the industry predict that up to one half of the ocean throughput will be intra-Asia trade using relatively small ships that travel relatively shorter distances, thereby reducing the demand for terminals while increasing the need for mid-stream sites. This issue will be taken up again in Chapter 8, where we address the question of whether a new container port in northeast Lantau should be built.

Notes

1. Assuming, of course, that options for building terminals 12, 13 and so on are not foreclosed.
2. See for example the *Asian Wall Street Journal*, 2 September 1996.
3. It is a well-known fact that normally growth in trade exceeds growth in GDP. Based on the relationship between trade and GDP growth, we adopt 1.2 as a conservative factor relating GDP to trade.

CHAPTER 8

To Build or Not to Build? Costs and Benefits of New Terminals

Hong Kong's Advantages and Disadvantages as a Container Port

As a container port Hong Kong's primary strengths include its strategic geographical location, outstanding facilities, efficient operations, and excellent supporting and related industries. Its weak-nesses include relatively shallow approach channels that do not meet the need of the newest generation of container ships, congestion at the Hong Kong-Shenzhen border checkpoints, congestion between the border and the container port, and high container handling fees such as tariffs to shipping lines and Terminal Handling Charges (THCs) to shippers.

Whereas geographical location and the depth of the port are natural conditions, the depth of the approach channels and quayside can be increased by dredging. Hong Kong is under pressure from shipping lines to prepare its port for increasingly larger container ships. To maintain its attractiveness to the shipping industry, the government has planned to dredge the Rambler Channel by mid-1997 to meet the needs of the largest ships.[1]

Other things being equal, efficient operations and excellent supporting industries are conducive to low costs and low prices for handling cargoes. However, in Hong Kong the capital costs (including land premium) and operating costs of terminal facilities are very high.[2] As a result, the tariffs charged by terminals and the

THCs are substantially higher than those by Singapore, South Korea and Taiwan.

To a large degree Hong Kong's port is a victim of the territory's own success. First, high land and labour costs are a consequence of the territory's booming economy. Second, there is congestions at the Hong Kong-Shenzhen border checkpoints and between the border and the container port have been caused by a large amount of China's container cargoes that pass through the port. Given the expected growth of such cargoes, the port will not be able to handle them effectively without support from its neighbouring ports in South China.

There is little doubt that Hong Kong will remain the most important hub port in South China, or even in all of China, for many years to come. But that does not necessarily mean that it must double or triple its existing ship berths in the next few years. From a regional perspective, a web of ports can be woven, with Hong Kong at its centre. If Yantian in Shenzhen were part of Hong Kong, then it would probably be the most natural location for Terminals 10, 11, and beyond; and the ports of Kwai Chung and Yantian would be regarded as two main components of the same hub port.

To Build or Not to Build Terminals 10 and 11?

Now that the controversy surrounding Terminal 9 has been resolved and that its construction may start in the last quarter of 1997, the next question is whether new terminals on Lantau (Terminals 10, 11, and beyond) should be built, and if so, when. In what follows, we first review the positions taken by the Port Development Board (PDB), terminal operators, shipping lines, and shippers. We then try to indicate how many new berths beyond Terminal 9 would be needed, and at what time. This is done by combining our estimates of Hong Kong's container handling capacity with alternative forecasts of future demand for its container handling services. As explained in Chapter 6, it would be very difficult to make predictions extending beyond ten years from now, so our analysis will be confined to the period 1997–2006.

Position of the Port Development Board

In its 1992 annual report the PDB "concluded that by 2011 the port will need 17 additional container berths each with a quay length of 320 metres" (p. 14) "over and above Terminal 9, which was expected to be due for completion by mid-1996" (p. 11). Appendix 22 of the same report (p. 56) indicates that from 1996 to 2001, seven new berths should be ready, to be followed by five additional berths by 2006, and another five by 2011. The same requirements are indicated again in Appendix 10 of the *1993/94 Annual Report*.

In its *1995 Annual Report*, the PDB revised the berth requirements upward. Taking into account the delay of Terminal 9, the report says that, "before the end of this decade a further 14 container berths will be required." Given the assumption that there would be four berths at Terminal 9, it implies that ten berths beyond Terminal 9 should be added by 2000, in contrast to the previous conclusion that only seven berths beyond Terminal 9 would be required by 2001.

In a leaflet entitled *The Port of Hong Kong*, the PDB claims that Terminals 9 and 10 should be ready by mid-1995 and mid-1997, respectively. Also, "Hong Kong will need, by the year 2006, at least 20 new container berths each having a capacity of 400,000 TEUs per year."[3] Thus, the total number of new berths to be available by 2006 increased by four from that given in the 1992 and 1993/1994 reports. Apparently, this reflected a more bullish view about future container cargoes as summarized in *Hong Kong Port Cargo Forecast 1995* (the "PDB Report" hereafter).

The PDB's view in the middle of 1996 was that Hong Kong faced a serious shortfall in berths due to the political wrangling over Terminal 9 and that new berths should be added as soon as possible, i.e., the first berth of Terminals 9, 10 and 11 should become operational by 1998,1999 and 2000, respectively.[4] Using the year 2000 as the point of reference, such a timetable would lead to slightly more berths than what is proposed in the 1992–94 Annual Reports but to about two to three berths less than what is proposed in the 1995 Annual Report. Most recently, in October 1996, the

Table 8.1
Estimates of Ship Berths Needed from 1996 to 2006

	1995	1997	1998	1999	2000	2001	2002	2003	2004	2005	2006	2011
Annual Report 1992, 1993/1994	1–2 (CT9)					4+7=11 (incl. of CT9)					4+12=16 (incl. of CT9)	4+17=21 (incl. of CT9)
Annual Report for the Period up to September 1995					4+10=14 (incl. of CT9)							
The Port of Hong Kong	1–2 (CT9)	5–6 (CT9 + CT10)	8 (CT9 + CT10)								20 (incl. of CT 9 & CT10)	
PDB Position as of October 1996				1 (CT9)	5 (CT9 + CT10)		9 (CT9, CT10 + CT11)					
Our Estimates if mid-stream is capped at 4m. TEUs by 1998					2 (CT9	4 (CT9)	6 (CT9 + CT10)	8 (CT9 + CT10)		10 (CT9, CT10+ CT11)	12 (CT9, CT10+ CT11)	
Our Estimates if mid-stream is allowed to expand to 6m. TEUs by 2006					2 (CT9)	4 (CT9)		6 (CT9 + CT10)	8 (CT9 + CT10)		10 (CT9, CT10 + CT11)	

PDB modified its position to require that the first berth of Terminals 10 and 11 be available in 2000 and 2002, respectively.[5]

The above PDB estimates of ship berth requirements are summarized in the first four rows of Table 8.1. Note that in its earlier reports, the PDB adopted 400,000 TEUs as the design capacity for a 320-metre berth but later raised the figure to 550,000 TEUs in view of substantial productivity increases at the container terminals. Given the actual throughput handled by the operators, there is no reason why the capacity cannot be raised, say, 700,000 TEUs per berth for the new terminals.

Despite a warning in the PDB Report (p. 12) that "by 1996 Hong Kong will have a shortfall of several container berths",[6] the agency also admits that "with the present terminal operators introducing new equipment and working methods, which have greatly improved the capacity of the existing eight terminals, this has not been a problem so far" (p. 7). Indeed, according to our analysis in Chapters 6 and 7, there was about 1.5 to 2.0 million TEUs of excess capacity in 1996.

As a government agency, the PDB aptly approaches the question of competition among regional ports largely from Hong Kong's perspective. As is evident in its 1993/1994 Annual Report, the agency was concerned that if demand was not met by adequate port facilities in Hong Kong, the demand would be lost forever.[7] There was also concern that a loss of cargo-handling would lead to a loss of supporting industries.

Positions of Different Terminal Operators

All existing and new terminal operators would agree that, regardless of the precision of the PDB forecasts, new container terminals are definitely needed to handle expanding container cargoes that move through Hong Kong and the neighbouring region. They have different views, however, about where new terminals should be built, i.e., on Lantau in Hong Kong, or in South China (for example, in Yantian, Shekou and Gaolan). A major point of uncertainty is that of how fast the South China ports can overcome weaknesses in customs procedures, regulation, and supporting services in order to make their cost advantage a powerful weapon of competition.

As a reflection of their own strategic positions and self interests, those Hong Kong operators that have greater stakes in South China ports are more likely to take a regional, as opposed to purely local, point of view. Their relative size also has a predictable effect on their concern about excess terminal capacity in Hong Kong. Specifically, the larger their share of container berths in Hong Kong, the more worried they would be about excess capacity. Small and new operators tend to be more supportive of building new berths on Lantau.

The differing attitudes of existing operators and new entrants to the cargo-handling industry are as expected. Faced with idle capacity, big operators see no need to hurry in building new berths in the territory, especially if they have similar facilities in China. Smaller and new operators, on the other hand, see new terminals as the only opportunities for them to increase their shares in a profitable market. Thus, while some existing operators warn that the high costs would make it difficult for investors to earn an acceptable rate of return, the new consortia appear eager to get into the container terminal business.

In Hong Kong, investment in terminals is made by private business, so the risk of significant over-capacity in the sense that new facilities will incur losses is small. However, the risk of excess competition in the sense of reducing the total industry profits below some socially optimal level cannot be ruled out.[8]

Positions of Shipping Lines and Shippers

Shippers constantly complain about Hong Kong's high THCs. Along with shipping lines, they have attributed high THCs in total or in part to what they perceive as monopoly pricing by the small number of terminal operators in the territory. Such complaints are to be expected given that terminal operators are direct or indirect customers of the terminals. If they do have a choice, they would naturally like to see more terminals and more independent terminal operators competing aggressively with one another for customers. Some see excess capacity as an effective way of forcing competition on the operators.

Pre-emptive Expansion of Container Terminals

Some small terminal operators and shippers argue that Hong Kong should build more berths to pre-empt the South China ports and to "lock in" Hong Kong's advantages from the economy of agglomeration. Let us examine the prospect of success for such a strategy. The container handling industry is very capital intensive and depends crucially on frequency of calls by ships as well as on a number of related services such as freight forwarding and insurance. There is no doubt about the economy of agglomeration and the disadvantages of being small. However, the real question is at what scale would such benefits become negligible or even negative. According to industry experts, many ports in the world operate efficiently at the level of 2 million TEUs. Excluding river trade and mid-stream operations, Hong Kong's total throughput was close to 8.3 million TEUs in 1995 and 8.6 million TEUs in 1996, more than four times the minimum efficiency level. Thus, it would not be very convincing to argue that the marginal benefits from further agglomeration would be substantial.

More importantly, in the future Yantian might build enough momentum to sustain itself despite the initial inertia. As a result, building new containers in Hong Kong with the objective of diluting Yantian's customer base will likely not succeed as a pre-emptive move. The recent history of excess capacity in the steel, ship-building, and computer chips industries has shown a clear pattern: low-cost producers push forward with expansion, despite excessive capacity at the global level, in the expectation that they will squeeze out high-cost competitors. From the point of view of the theory of strategy, there are few justifications for a high-cost port like Hong Kong to expand its facilities for the purpose of pre-emption. As Yantian moves forward undauntedly with its own expansion plan, new terminals in Hong Kong built primarily out of a pre-emptive motive would likely become a losing proposition.

Hong Kong's Overall Perspective

Conflict of Interests

As noted, the interests of shippers are in conflict with those of shipping lines, and the interests of shipping lines are in conflict with those of major existing terminal operators. Such conflicts of interest, however, are not peculiar to the container handling industry. Shippers are customers of shipping lines which in turn are customers of terminals. Terminal operators (along with mid-stream operators) provide an "intermediate input service" to downstream producers like shipping lines, which in turn provide a final shipping service to shippers. As in any other industry, customers prefer low prices while suppliers attempt to choose prices to maximize their profits.

We have explained why the major terminal operators' interests are not identical to those of the new and small operators. That the latter should take a more affirmative position on new terminals than that taken by the former is to be expected. The marginal revenue of throughput is inversely related to an operator's current market share, which is positively correlated with his relative capacity. In addition, small operators can benefit from additional capacity in light of the need to serve many ships from a single consortium.

Shippers, shipping lines and terminal operators all depend on one another for profits and even survival, but their relative market power determines the distribution of profits among and within the three different groups. If tariffs are lower (whether because there are more independent terminal operators or because there is greater excess capacity), there is more surplus to be divided between shipping lines and shippers. Similarly, if shipping costs (THCs plus freight rates) are lower, there is more surplus to be shared among manufacturers, traders, retailers and final consumers.

Competitive, Monopoly, Oligopoly, and Socially Optimal Tariffs

Competitive prices are equal to both marginal costs and average cost so that there are no excess profits. Monopoly prices are those prices that maximize total industry profits. From the point of view of shippers and shipping lines, competitive tariffs charged by the terminal operators are the best that the former groups can hope for. But from the point of view of terminal operators as a whole, they would have an incentive to set monopoly tariffs through some ideal cartel arrangement.

In reality, tariffs are set by terminal operators through oligopolistic competition in recognition of their interdependence. Thus, unless they have a cartel arrangement with side payments, the actual tariffs would be somewhere between competitive tariffs and monopoly tariffs. Since there are only two independent, major terminal operators in Hong Kong, it would be easier for them to find out which one of them has cheated on an agreement than if there were more operators, especially when the shipping lines are organized into large consortia. Thus, agreements about tariffs and market shares would hold up better, and one would expect oligopoly tariffs to be closer to monopoly tariffs than to competitive tariffs. Tariffs are used here to embrace volume discounts, services and other aspects of the pricing package that terminal operators charge shipping lines. In the following discussion we shall use the industry's monopoly tariffs as an approximation for oligopoly tariffs.

From the point of view of Hong Kong's economy as a whole, what tariffs are optimal in the sense of maximizing Hong Kong's aggregate economic welfare, say Gross Domestic Product (GDP)? Strictly from the point of view of producing goods for export and importing goods for consumption and investment, perfectly competitive (terminal) tariffs will be called for, because any deviation from competitive tariffs would create distortions that would reduce GDP. Analogous to the standard analysis of taxes on imports, one can show that if the terminal tariffs are above their competitive levels, the total loss in (a) consumer's surplus from reduced imports for consumption, and (b) producer surplus from reduced exports and imports, will be greater than the gain in profits by the terminal operators.

However, that is not the entire story, because Hong Kong is an entrepôt by virtue of its strategic location on the world's shipping lanes. From an operational point of view, the port is said to enjoy some monopoly power so long as an increase in its total handling charges on services provided in the territory for entrepôt cargoes (i.e., charges for container handling, insurance, freight forwarding, etc.) has only a finite impact on the quantity of cargoes passing through it. In other words demand will not disappear when the total charges are raised by a small amount. In economic jargon, demand for handling entrepôt cargoes is "finitely elastic" but not infinitely elastic with respect to total handling charges.

The fact that Hong Kong is a key entrepôt implies that its services are not readily substitutable by other sea ports, so it must enjoy certain monopoly power in handling cargoes for other economies. When Hong Kong increases total charges for services provided in the territory (including the terminals' tariffs), it will lose some business; but foreign consumers, producers and intermediate service providers would also be forced to absorb part of that increase.

A monopoly price of Hong Kong's entrepôt trade services would maximize its total surplus from handling entrepôt cargoes. Because there is no substitution between container handling and other complementary services,[9] setting the monopoly price for all entrepôt trade services is equivalent to setting monopoly tariffs subject to given prices for all other services.

To maximize Hong Kong's aggregate economic benefits, the "socially optimal tariffs" would comprise two parts: monopoly tariffs for entrepôt cargoes but competitive tariffs for Hong Kong's own exports and imports. If for whatever reasons only one uniform set of tariffs can be charged, then there would be a trade-off or compromise between the competitive and monopoly levels. Furthermore, the uniform optimal tariffs will be closer to monopoly tariffs if entrepôt trade becomes relatively more important vis-a-vis Hong Kong's own cargoes. When the fact that some of the entrepôt cargoes are from companies owned by Hong Kong investors is taken into account, the socially optimal tariffs should be adjusted downward but never to the fully competitive level.

From standard economic analysis, we know that any profit-maximizing price is inversely related to the "price elasticity of demand" (i.e., the percentage reduction in demand for Hong Kong's container handling services in response to a one percentage point increase in the tariffs). Thus, one would expect the monopoly tariffs on entrepôt cargoes to be lower than are those for the territory's own cargoes, as the former have more alternatives than do the latter to using the port of Hong Kong. To the extent that the terminal operators can set different tariffs for entrepôt cargoes and the territory's own cargoes, the tariffs on the former would be close to the socially optimal level, but the tariffs on the latter (i.e. Hong Kong's own imports and exports) would be much too high. The latter tariffs would exceed not only the socially optimal tariffs for the territory's own cargoes (i.e. the competitive tariffs), but also the monopoly tariffs on entrepôt cargoes.

In reality, price discrimination on the basis of cargo source and destination is not possible because of prohibitive monitoring costs; therefore, the joint profit-maximizing tariffs for all cargoes will lie between the two sets of individually profit-maximizing tariffs. Under these circumstances, profit-maximizing tariffs for the territory's own cargoes exceed the joint-profit-maximizing tariffs, which in turn exceed the profit-maximizing tariffs for entrepôt cargoes. Upon comparison with the socially optimal tariffs, it follows that a small reduction in tariffs on all cargoes improves aggregate economic welfare.

Let us summarize our findings from the above analysis. Given that there are only two independent, major terminal operators in Hong Kong, one would expect the oligopoly tariffs to be closer to their joint profit-maxmizing (cartel) tariffs than to the competitive tariffs. If different tariffs can be set according to the source and destination of cargoes, then the tariffs for entrepôt cargoes would be close to the socially optimal levels, but the tariffs for the territory's own cargoes would be too high. In practice, uniform tariffs must be set regardless of the cargoes' source and destination; therefore, the tariffs would be too high for both types of cargoes. In this last case, a reduction in tariffs would improve economic welfare in both dimensions.

When demand for capacity increases, the most efficient way to lower the oligopoly tariffs would be to increase the number of independent terminal operators, just as in the case of the telecommunications industry.

Local versus Regional Perspectives

Tension exists between efficient regional location and co-ordination of port facilities on one hand, and a desire to "lock in" demand for Hong Kong's port services by expanding the territory's port facilities on the other. Like the relocation of labour-intensive assembly processes from the territory to South China, the relocation of land-intensive and labour-intensive components of container handling to China will likely increase the territory's gross national product (GNP) and perhaps even its gross domestic product (GDP), if the resources released can be put to better use.

Thus, it is important to step back and take a regional rather than a purely local perspective in assessing Hong Kong's advantages and disadvantages vis-a-vis the South China ports, and to explore co-operation that can lead to win-win situations. Some people in the industry suggest that the port of Hong Kong should concentrate on large and very large ships, while smaller ships should better be handled in such Chinese ports as Yantian and Shekou. Although this is one possible division of labour, the logic is not immediately obvious, because Yantian can meet the depth requirements of the largest container ships even better than Hong Kong can. Perhaps a division based on functions (container handling operations such as cargo consolidation, yard and quay operations management, information processing, warehousing and supporting services such as insurance and freight forwarding) would be more consistent with the principle of comparative advantage than would be a division based on ship sizes.

The above-proposed division of labour in port cargo handling is also consistent with Hong Kong's own experience in relocating product-assembly activities to the Pearl River Delta. Some people might raise concerns about the possibility of "hollowing out": fearing that once container handling is lost other related services will be gone too. However, the experience of other ports seems to suggest that much of the administrative, commercial and financial activity related to the movement of container cargoes (e.g., trading, insurance, trade financing, freight brokering forwarding, and ship brokering) could remain in Hong Kong even though the actual loading, unloading, stacking, consolidation and warehousing of containers are done more cost-effectively in the other ports.

Infrastructure and Environment

To assess the economic benefits and costs of building additional terminals, the government should calculate the additional costs in public infrastructure investment needed to support the new terminals. The cost of the Port Rail Link is a case in point. Besides weighing the incremental economic benefits and costs associated with the building of new terminals, environmental costs and social costs should also be brought to bear before the optimal number of berths is decided upon.

According to experts in the industry, about 30,000 truck trips each day are required to serve the needs of the existing container port. According to the PDB's forecast, the volume of ocean containers in 2006 would be about two and half times that in 1995. The projection implies that there would be more than double the amount of current truck trips per day, unless a large portion of the increased cargoes between the container port and China are moved by rail. In addition to dealing with more tractor-trailers, Hong Kong's environment may also suffer from the "littering" of empty containers in the New Territories.

On the basis of cost and benefit calculations, the government should determine whether it is desirable to meet the entire increase in demand or just to meet a part of it. After some critical level of container handling is reached, the marginal social costs of providing additional berths could rise steeply such that they would exceed the marginal social benefits. In any event, it is not clear why Hong Kong should, as implied by the PDB Report, meet alone whatever demand there would be for its container handling facilities.[10]

Supply Follows Forecast Demand: The Timing of Terminals 10 and 11

In this section, we shall follow the same approach as does the PDB in determining the need for Terminals 10 and 11 within the ten-year period leading up to 2006. Juxtaposing the forecasted demand for handling ocean cargoes against the estimated handling capacity of Terminals 1 to 9, we can obtain some ideas about the timing of these two new terminals before infrastructural and environmental considerations are brought to bear.

The results obtained from a comparison of Table 7.1 in Chapter 7 and Table 6.2 and Table 6.3 in Chapter 6 are presented in the bottom two rows of Table 8.1. The figures in the fifth row are based on the assumption that mid-stream operations will be capped at 4 million TEUs from 1998 onwards. Under this assumption, Terminal 10 would be needed in 2002, while Terminal 11 would be needed in 2005. If mid-stream operations are allowed to expand gradually from 4 million TEUs in 1998 to 6 million TEUs by 2006, then the construction of both terminals can be delayed for one year, as indicated in the last row. The total capacity under these two scenarios ("expansion" and "capped") are given in Tables 8.2 and 8.3 respectively. These estimates of capacity against the forecasted ocean container throughput are presented graphically in Figures 8.1 and 8.2.

The government should realize that demand can never be exactly predicted and should therefore consider the costs of excess capacity against the costs of insufficient capacity. Thus far, mid-

Table 8.2
Future Capacity of Terminals & Mid-Stream Operations
("Expansion" Projection)

		Million TEUs in					
		1996	1998	2000	2003	2004	2006
(1)	CT 1–9 & Mid-stream	13.5 to 14.0	16.0	18.0	20.0 to 20.4	20.3 to 21.0	20.8 to 21.5
(2)	CT 10						
	(2 berths)				1.4	1.4	1.4
	(2 more)					1.4	1.4
(3)	CT 11						
	(2 by 2006)						1.4
(4)	Total Capacity (beyond 2000)				21.4 to 21.8	23.1 to 23.8	25.0 to 25.7

Source: Item (1) from Table 6.2, data for Year 2003 intepolated

Note: Item (4), the total capacity, is used in Figure 8.1 (top curves) showing the consequences of allowing mid-stream operations to "expand" gradually.

Table 8.3
Future Capacity of Terminals & Mid-Stream Operations
("Capped" Projection)

		Million TEUs in					
		1996	1998	2000	2002	2004	2006
(5)	CT 1–9 & Mid-stream	13.5 to 14.0	16.0	17.5	18.8	18.8 to 19.5	18.8 to 19.5
(6)	CT 10						
	(2 berths)				1.4	1.4	1.4
	(2 more)					1.4	1.4
(7)	CT 11						
	(all 4 by 2006)						2.8
(8)	Total Capacity (beyond 2000)				20.2	21.6 to 22.3	24.4 to 25.1

Source: Item (5) from Table 6.3

Note: Item (8) is used in Figure 8.2 (top curves) showing the consequences of "capped" capacity of mid-stream operations by 1998 at 4 million TEUs.

Figure 8.1
Total Capacity of Terminals & Mid-Stream Operations, 1996–2006
("Expansion" Projection)

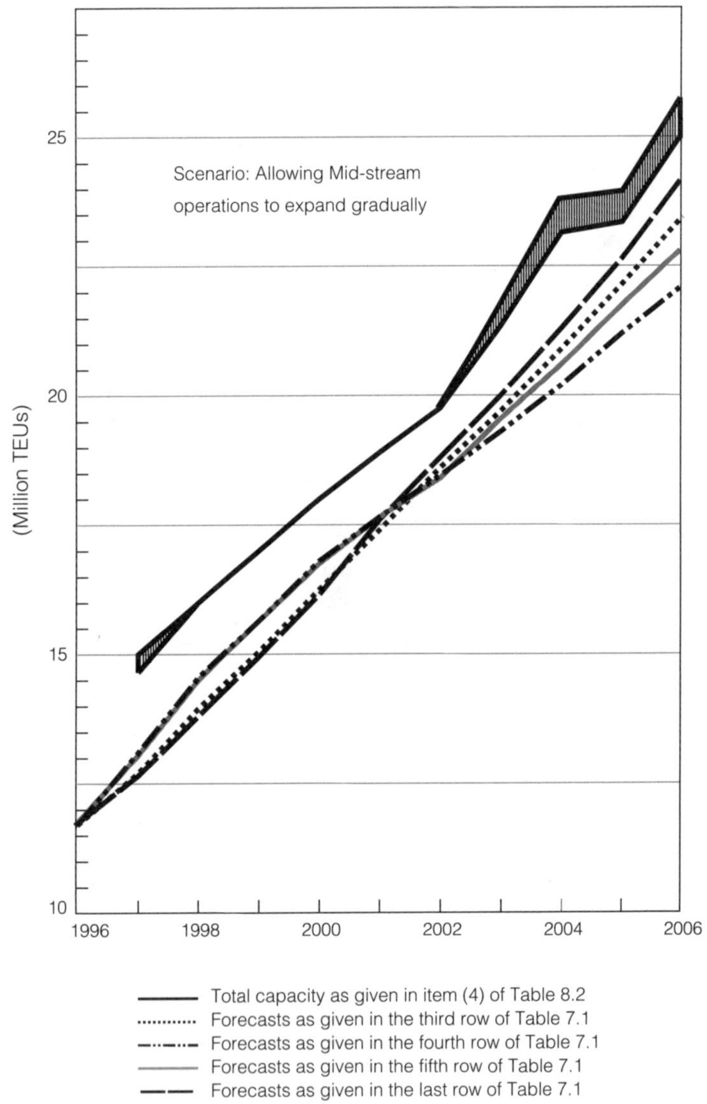

Figure 8.2
Total Capacity of Terminals & Mid-Stream Operations, 1996–2006
("Capped" Projection)

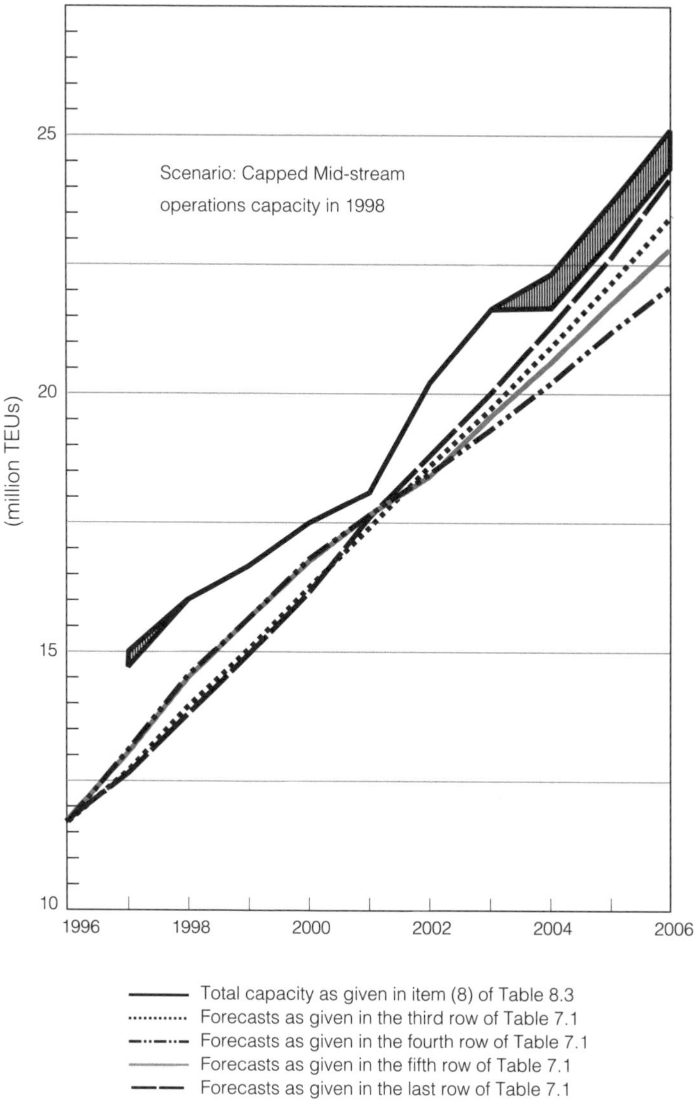

stream operations have served as a safety valve capable of adjusting to unforeseen changes in demand. It would seem that mid-stream operations, like a stand-by generator to meet unforeseen gaps between the normal demand and supply of electricity, provide a more cost-effective means of dealing with unexpected demand than would excessive capacity at the terminals.

As can be seen from Figures 8.1 and 8.2, the estimates of capacity have incorporated some excess capacity throughout the entire period from 1996 to 2006. Nevertheless, in cognizance of unexpected changes in demand (which may go in either direction), we suppose that the needed berths should be constructed one year earlier to provide an even larger safety margin in avoiding inadequate capacity in case demand proves to be stronger than expected. Terminals 10 and 11 should then be available for service by 2001 and 2004, respectively. An important finding of our analysis is that our conservative results do not support the PDB's position, that Terminals 10 and 11 should be brought on stream in 2000 and 2002, respectively, or earlier.

The differences between our findings and those of the PDB come from two sources. First, the PDB has been quite optimistic about future container throughput. In its earlier Port Cargo forecasts, the PDB erred by underestimating the growth in container throughput. So, in its 1995 forecasting exercise, PDB made an effort to avoid making a similar mistake by being more bullish. Unfortunately, economic forecasting is often at the mercy of unexpected turning points. This time around, the PDB might have erred by being too optimistic.

Second, the PDB has used design capacity figures that are well below those that can be achieved by the operators. PDB's current design capacity figure of 550,000 TEUs per berth is substantially below the best practice of one million TEUs achieved by one operator, Sea-Land Orient Terminals (SLOT), in 1996. In our own calculations, we used 666,700 TEUs per berth for Terminals 1 to 8. Since Terminal 9 has six berths, we assume the total capacity will reach a ceiling of 3.5 million TEUs four years after it first becomes operational, and we assume 700,000 TEUs per berth for the new subsequent terminals.

Quantitatively speaking, the above differences in the requirements of new container berths are accounted for primarily by differences in the estimates of capacity. Capacity estimates are secondarily only affected by differences in demand forecasts. To the extent that the PDB's adopted design capacity figures do not reflect reality, the "trigger point mechanism" and the planning process would lead to systematic errors.

Is There a Need for New Terminals Beyond Terminal 9?

Yes to Terminals 10 and 11

On the basis of our analysis of expected demand for container handling services in the next ten years and the capacity of Terminals 1 to 9, we conclude that Terminals 10 and 11 would be needed by 2002–2003 and 2005–2006, respectively. Our difference with the PDB lies in the timing of constructing the terminals, not in the idea that they are needed.

Some critics of the PDB have gone so far as to argue that it is not necessary to build new terminals beyond Terminal 9 or even to build Terminal 9 itself. One of their key arguments concerns the adverse environmental impact of the new terminals. Another key argument is that the South China ports such as Yantian and Gaolan have enough physical space to build new terminals so that the total demand for container handling services in South China can be met without building any new berths in Hong Kong.

We do not have the expertise to critically assess the environmental argument as a reason for not building any new container berths. We shall concentrate on the second argument instead. As explained in Chapter 6, given the uncertainty about the realized capacity of the South China ports, it would not be very useful to add up their maximum engineering capacities, and then juxtapose them against the forecasted demand for the region, in order to determine whether any new berths are needed in Hong Kong. It suffices to repeat a point already made before, that cargoes and ships do not go to the ports simply because ports have the handling capacity.

Even if in twenty years Hong Kong's terminals cannot compete price-wise against Yantian and Gaolan, Hong Kong can and should still build new terminals and make handsome profits before the latter ports develop into serious competitors. Creative destruction is the name of the game under capitalism: while no firm can expect to be a leader forever, a leading firm should try to collect as much monopoly profit during the finite period of its dominance. For high-technology industries, a window of opportunity may be open for just a few years, but that does not deter high-tech firms from investing huge sums of money to establish their temporary monopoly positions. In other words, if new terminals in Hong Kong can more than recover their investment before Yantian develops into a fierce competitor, there is no reason not to reap the temporary profits.

Terminals 12, 13 and Beyond

Given that it is inherently difficult to predict the growth of China trade via Hong Kong, or to predict the competitiveness of the South China ports, and the port of Kaohsiung in Taiwan, more than a decade from now, we cannot say that Terminals 12 and 13 would be needed with the same degree of conviction that we can say it now about Terminals 10 and 11. Fortunately, there is no need to make any irreversible decision about Terminals 10 and 11 for another five years.

With the unfolding of information about demand for Hong Kong's container handling facilities and rival supply by regional competitors, a rational decision can be made later. For now, the planners of the territory's port facilities should ensure that the possibility of building beyond Terminal 11 is not foreclosed, but at the same time they should be aware that the scale of the new container port could be much smaller than what was once envisioned. Flexibility in the planning of such facilities is thus essential if scarce resources are not to be wasted. However, to the extent that a large investment in infrastructure specific to the new container port is an irreversible fixed cost, a smaller port may translate into an increase in the cost of insufficient demand relative to the cost of insufficient capacity. Other things being the same, this consideration will weigh against a large planned excess capacity for the purpose of maintaining a large safety margin.

The Trigger Point Mechanism and Competition

To protect the financial interests of the private investors who developed the terminals, the PDB has adopted a "trigger point mechanism" to ensure that the government will not initiate the development of new terminals unless and until forecasted demand five years ahead equals existing capacity by a sufficient margin. As a consequence it is possible that supply lags behind demand, in contrast to the case of Singapore where the government builds terminals ahead of demand.

It has been pointed out earlier by the Hong Kong Centre for Economic Research (1992) that "since the amount of excess demand is determined in part by the level of handling charges set by terminal operators, this mechanism provides existing operators with a tool to limit competition by reducing the rate of entry." In other words, oligopoly tariffs tend to lower the realized cargo volumes. If the industry becomes appropriately competitive with the addition of new operators, and if anti-competitive trade practices are prohibited or discouraged by law, then the drawback of the trigger point mechanism would disappear.

Thus, a better way to deal with the drawback of the trigger point mechanism is to increase competition directly by increasing the number of independent terminal operators. Creating excess capacity to bring about more competition, without changing the industry structure, is an indirect and thus inefficient way to achieve the objective. Cartel-type agreements are more easily enforceable with a very small number of operators but more difficult to enforce if the number of independent operators is sufficiently large. With a small number of independent operators, an increase in capacity would likely lead to an increase in idle capacity rather than enhancing the competitive nature of the industry.

In contrast, with more independent operators, an increase in capacity would likely result in greater competition. But with an appropriate competitive environment, there is no need to consider excessive capacity at all. Furthermore, using excess capacity to lower tariffs without changing the industry structure is an inefficient method because it results in wasteful idle capactiy.

Notes

1. According to industry experts, the fifth-generation container ships have an average capacity of about 6000 TEUs, 20% to 30% more than the fourth generation ships have. When fully loaded, these newer ships draw about fifteen metres of water. However, the depth of Hong Kong's harbour ranges only from 12.5 to 14.5 metres. If dredging is not done, the large ships would be forced to carry less than their full load, as they currently do, thus increasing the cost of calling the port of Hong Kong.
2. According to an industry expert, construction and equipment costs for a 320-metre berth with fifteen hectares of land in Hong Kong is currently about $2 billion. Adding the land premium would push the total capital cost even higher, to $2.5 to $3 billion per berth. In contrast, the cost of new ports in Guangdong, at about $1.1 billion, is about 50% less than the cost of new ports in Hong Kong. The operating cost per TEU in Hong Kong is $350, while that in South China is about $300.
3. A similar statement appeared in another leaflet entitled "What is PADS?"
4. *Hong Kong Economic Journal*, 15 June 1996.
5. *Ming Pao*, 10 October 1996.
6. The following statements were made in the PDB's *Annual Report 1993/1994* (5). "The Hong Kong Government has estimated that a two-year delay in building Terminal 9 will cost the Hong Kong economy some $20 billion over the decade from 1997 to 2006. Some private sector economists regard this figure as conservative." According to the *Hong Kong Economic Journal*, 15 June 1996, the PDB claimed that if Terminals 10 and 11 were not built, Hong Kong would lose HK$78 billion from 1997 to 2011.

7. For example, in the PDB's *Annual Report 1993/1994* (p. 7) it was stated that, "once lost to other ports, such business would be very difficult to win back."
8. If the new terminals (e.g., Terminals 10 and 11) on Lantau are less profitable, then lower land premium will result in the bidding process. Even though the government is not necessarily concerned about the premium that it collects, whether the land is given to the best use is still a relevant consideration.
9. In economics jargon, the requirement of container handling and the requirement of other services are subject to a technology of "fixed proportions", so that changes in the relative prices of the two types of inputs have no effect on the optimal choice of inputs.
10. On page 1.2 of the PDB's *Port Cargo Forecast 1995*, it is stated that, "the next stage of the PDSR is to formulate a port development plan and programme in order to match supply of port infrastructure with the forecast demand for its use."

CHAPTER 9

Conclusions and Recommendations

The port facilities and container handling services in Hong Kong have done very well in the last several decades to serve the needs of the territory and have maintained their leading position in the world over the last decade. The rapid expansion of service is due primarily to the opening of China and the expansion of world and regional trade. The efficiency and adaptability of terminal and mid-stream operators in meeting this expansion is in itself an exemplary feat. The advantage of an institutional arrangement that relies on private initiatives in the development and operation of port facilities and container handling services is clearly evident. Hong Kong is a pioneer in this regard. The government and the industry deserve credits for making such an arrangement viable.

While the future of this sector is expected to remain robust, the changing landscape of trade and investment flows, the emergence of new ports in the region, and the rising cost structure in Hong Kong will bring fresh challenges. In the previous chapters we analyzed the potential competitors in the region, estimated Hong Kong's handling capacity, and generated forecasts of its container throughput. Up-to-date forecasts about the growth prospect of Hong Kong's major trading partners have been used in this book. We concluded that while South China ports such as Yantian in Shenzhen could potentially pose serious challenges to Hong Kong's dominance, it would likely take ten years or more before they fully realize their potential. In comparison, the threat of port Kaohsiung in Taiwan could be more immediate. Its impact would be primarily upon Hong Kong's role in the transshipment of China cargoes; and the timing of this impact would depend on the political relationship between Mainland China and Taiwan.

We estimated the handling capacity of Terminals 1 to 9 and mid-stream operations from 1996 to 2006. The first berth of Terminal 9 is expected to become available in first quarter of year 2000, and the total capacity of Terminals 1 to 9 is expected to increase from about 10 million TEUs in 1996 to 13.5 million TEUs by 2000, and may reach 15.5 million TEUs by 2004. If the capacity of mid-stream operations is allowed to grow from 3.5 million TEUs in 1996 to 4 million TEUs by 1998 and to stay at that level thereafter, then the total capacity of Terminals 1 to 9 and mid-stream operations is expected to rise from 13.5 to 14 million TEUs in 1996 to 17.5 million TEUs in 2000, and may reach 19.5 million TEUs by 2004. If, instead, the capacity of mid-stream operations is allowed to rise by 0.5 million TEUs every two years beginning with 3.5 million TEUs in 1996, then the total capacity of Terminals 1 to 9 and mid-stream operations would rise from 13.5 to 14 million TEUs in 1996 to 18 million TEUs by 2000, and may rise to 21.5 million TEUs by 2006.

Based on the actual throughput in 1995 and 1996 and the revised growth trend of ocean container throughput, we generated three sets of forecasts from 1997 to 2006. Upon comparison with the original Port Development Board (PDB) forecasts, we discovered that the PDB estimates of the ocean container throughput are higher than ours by over 1 million TEUs in 1997, and by 1.5 to 2.8 million TEUs by 2006. The alternative forecasts came out to be about 13 million TEUs in 1997 and 22 to 23.5 million TEUs by 2006. That is to say, despite an expected slow-down in the growth of ocean container throughput compared with figures over the last decade, Hong Kong needs additional container terminals in order to meet the forecasted demand for the territory's container handling services.

Based on the above estimates and following the PDB's trigger point mechanism, we conclude that Terminals 10 and 11 will be needed in 2002 and 2005, respectively, if the capacity of mid-stream operations is capped at 4 million TEUs. However, the two new terminals can be delayed for one year if mid-stream operations are allowed to increase beyond 4 million TEUs by 0.5 million TEUs every two years. Thus, our findings do not support PDB's conclusions that these two new terminals should come on stream sooner — in 2000 and 2002, respectively, or even earlier. We have also argued that in making decisions about building additional terminals, the government should consider their marginal environmental and social costs along with their marginal economic costs. The incorporation of these costs would likely reduce the optimal size of Hong Kong's container port relative to that dictated solely by demand.

Regardless of the exact timing of the construction of Terminals 10 and 11, we recommend that maximum flexibility be maintained so that additional terminals would not be foreclosed. At the same time, the government should also be prepared for the possibility that no terminals beyond Terminal 11 may be built, given the possible changes in the external economic environment and Hong Kong's own comparative advantage in container handling.

Hong Kong's container handling industry is an oligopoly with plenty of room for co-operation. Due to (a) the fact that there are only two independent, major terminal operators, (b) a lack of legal restraints on anti-competitive behaviour such as price setting and non-poaching agreements, and (c) corroborative evidence of oligopolistic co-operation such as non-poaching agreements, there is possible scope for enhancing competition in the sector. One can argue that the outcome of the operators' oligopolistic competition would not be too different from that of a single operator or a cartel of operators. Even if a more conservative position is taken, one cannot escape the conclusion that the industry's competition can be enhanced by the presence of additional independent operators.

To the extent that competition in the container handling industry is inadequate, the interests of shipping lines and shippers (manufacturers, traders and buyers) are adversely affected. However, competition that would bring terminal charges down to marginal costs and average costs is not in Hong Kong's overall economic interest, either. As a strategically located entrepôt, Hong Kong can take advantage of its strategic entrepôt position in handling containers for other economies. Thus, a balance between the economic benefits derived from local cargoes and those derived from entrepôt cargoes is needed.

If terminal operators are able to set different tariffs for entrepôt cargoes on the one hand and for the territory's own cargoes on the other, entrepôt tariffs would not be much below those that are socially optimal for Hong Kong; in that case, however tariffs for local cargoes would be much too high. Increased competition would improve Hong Kong's economic welfare derived from its own exports and imports, but it would be partially offset by a loss in economic welfare derived from entrepôt cargoes.

In practice, operators are not able to set different tariffs for the territory's own cargoes and entrepôt cargoes. They would have to make a trade-off between the two sources of profits. The resulting price is between the individually profit-maximizing tariffs, but that price is socially too high for both types of cargoes. In this case, a small reduction in terminal tariffs will increase Hong Kong's total economic welfare on both counts. However, at the socially optimal point, there would be a trade-off between the economic welfare derived from the two kinds of cargoes.

If discriminatory policies are possible, then the Hong Kong government can find a way to lower the terminal tariffs borne by the territory's own cargoes. This can be done by a subsidy for local cargoes to be financed by a general tax on all cargoes. If discriminatory policies are ruled out due to international commitments or to Hong Kong's own principles in setting economic policies, then increasing competition among terminal operators to bring about lower tariffs would be desirable. Given the trade-off discussed above, in no case should the government bring about excessive competition, one that will force terminal tariffs down to their average and marginal costs.

On the basis of the above findings, we make three recommendations: First, we recommend that new terminal operators be added to the industry in the future.

At present, legislation against unfair and anti-competitive practices does not exist. However, legislation for the purpose of enhancing competition should be applied economy-wide rather than Industry-specific. How to enforce such legislation is beyond the scope of our study. In any event, we believe that, for the purpose of increasing competition in Hong Kong's container handling industry, increasing the number of independent operators would have the greatest effect while the role of legislation would only have a secondary effect.

Let us restate our conclusion concerning the potential drawback of the PDB's trigger point mechanism as a way of determining the need for new terminals. If there is insufficient competition among terminal operators, like it is now, then tariffs will be too high and the realized demand for container handling services will be too low, leading to under provision of terminals. If the degree of competition among terminal operators is increased to an appropriate level, then tariffs will be socially optimal. In that case, the demand would be met, and the shortage of terminals would disappear.

Second, we do not recommend the use of excess capacity to bring about more competition because the use of excess capacity is inefficient and thus not cost effective. Cartel-type agreements are more easily enforceable with a very small number of operators but more difficult to enforce if the number of independent operators is relatively large. If the number of independent operators remains small, then an increase in capacity would likely lead to an increase in idle capacity rather than to a more competitive market structure. While tariffs may come down as a result, it would not be an optimal outcome. In contrast, if there are more independent operators, an increase in capacity would likely result in increased competition. But if the industry is appropriately competitive, there is no need to consider excess capacity as a remedial measure to meet demand.

Finally, we recommend that marginal environmental and social costs along with economic costs should be considered in determining the desirability of new container terminals.

Appendix

Table 2.1

Freight Movement through the Port of Hong Kong by Modes of Transport 1983–1995

('000 tonnes)

Year	1983	1984	1985	1986	1987	1988	1989	1990	1991	1992#	1993	1994	1995
Inward Cargo													
Ocean	24138	26451	29657	35101	38942	44258	45792	46242	52899	58923	68226	76672	87048
Ocean Growth Rate (%)		9.58	12.12	18.36	10.94	13.65	3.47	0.98	14.40	11.39	15.79	12.38	13.53
River	4244	4258	5264	6315	6152	6009	5477	6026	6722	11627	11783	16172	14723
River Growth Rate (%)		0.33	23.63	19.95	–2.58	–2.32	–8.85	10.02	11.56	72.97	1.34	37.25	–8.96
Ocean and River	28382	30709	34921	41416	45094	50267	51269	52268	59621	70550	80010	92844	101771
Ocean and River Growth Rate (%)		8.20	13.71	18.60	8.88	11.47	1.99	1.95	14.07	18.33	13.41	16.04	9.61
Ocean/(Ocean + River) (%)	85.05	86.13	84.93	84.75	86.36	88.05	89.32	88.47	88.72	83.52	85.27	82.58	85.53
All Modes of Transport †	30655	33543	38107	45741	50377	55996	57595	59208	67400	79093	89342	102265	111716*
(Ocean + River) / All Modes of Transport (%)	92.59	91.55	91.64	90.54	89.51	89.77	89.02	88.28	88.46	89.20	89.55	90.79	91.10
Outward Cargo													
Ocean	7430	8842	10032	12367	14615	17063	18863	19766	23546	24524	27873	34274	40127
Ocean Growth Rate (%)		19.00	13.46	23.28	18.17	16.75	10.55	4.79	19.13	4.15	13.66	22.97	17.08
River	1174	1682	2600	2506	3258	4060	3549	3262	4425	7706	10255	13907	14009
River Growth Rate (%)		43.27	54.59	–3.63	30.00	24.63	–12.58	–8.11	35.67	74.14	33.08	35.61	0.74
Ocean and River	8604	10524	12633	14873	17873	21123	22412	23028	27971	32229	38128	48181	54136

Ocean and River Growth Rate (%)		22.32	20.04	17.74	20.17	18.18	6.10	2.75	21.47	15.22	18.30	26.37	12.36
Ocean/(Ocean + River) (%)	86.36	84.02	79.42	83.15	81.77	80.78	84.16	85.84	84.18	76.09	73.10	71.14	74.12
All Modes of Transport	9710	11943	14332	17275	21064	25231	27145	28340	34280	38869	44809	54824	61160
(Ocean + River) / All Modes of Transport (%)	88.61	88.12	88.14	86.10	84.85	83.72	82.56	81.26	81.60	82.92	85.09	87.88	88.52
Inward + Outward Cargo													
Ocean	31569	35293	39689	47469	53557	61321	64655	66008	76445	83446	96100	110947	127175
Ocean Growth Rate (%)		11.80	12.46	19.60	12.83	14.50	5.44	2.09	15.81	9.16	15.16	15.45	14.63
River	5418	5940	7864	8820	9409	10069	9027	9287	11147	19333	22038	30079	28732
River Growth Rate (%)		9.63	32.40	12.16	6.68	7.01	−10.36	2.89	20.03	73.43	13.99	36.48	−4.46
Ocean and River	36987	41233	47553	56289	62867	71390	73681	75295	87592	102760	118138	141025	155907
Ocean and River Growth Rate (%)		11.46	15.33	18.37	11.86	13.35	3.21	2.19	16.33	17.34	14.94	19.37	10.55
Ocean/(Ocean + River) (%)	85.35	85.59	83.46	84.33	85.06	85.90	87.75	87.67	87.27	81.19	81.35	78.67	81.57
All Modes of Transport	40365	45486	52440	63016	71441	81227	84741	87547	101680	117962	134151	157089	172876
(Ocean + River) / All Modes of Transport (%)	91.63	90.65	90.68	89.32	88.14	87.89	86.95	86.01	86.15	87.13	88.06	89.77	90.18

Sources:
Hong Kong Port Development Board, *Port Cargo Forecasts*, December 1993; February 1996, Table A1.
Hong Kong Census and Statistics Department, *Hong Kong Shipping Statistics*, January – March 1996, Table E1.

Note:
Inward cargo figures exclude gravel & crushed stone imported by conveyor transport system across the border.
\# New definitions have been adopted for ocean and river cargoes since 1992.
* Livestock was excluded in 1995 data.
† The modes of transport include ocean, river, rail, road, and air.

Table 2.2
Direct Ocean Traffic by Cargo Type 1987–1994 ('000 tonnes)

Cargo Type	1987	1988	1989	1990	1991	1992#	1993	1994
Inward								
Break bulk	3139	3200	2922	2946	3950	4183	4653	4613
% share	9.55	8.54	7.45	7.33	8.74	8.25	7.97	7.34
Growth Rate (%)		1.95	-8.68	0.81	34.10	5.91	11.23	–0.86
Dry bulk	15417	17536	17293	16914	16848	17362	18585	17034
% share	46.89	46.82	44.07	42.11	37.26	34.23	31.84	27.11
Growth Rate (%)		13.74	–1.39	–2.19	–0.38	3.05	7.04	–8.35
Liquid bulk	5847	6558	8268	7890	8949	10882	13001	13968
% share	17.78	17.51	21.07	19.64	19.79	21.45	22.27	22.26
Growth rate		12.15	26.05	–4.55	13.42	21.60	19.47	7.60
Containerized cargo	84.75	10158	10758	12418	15469	18301	22139	27206
% share	25.78	27.12	27.42	30.92	34.21	36.08	37.92	43.29
Growth rate		19.66	5.90	15.44	24.56	18.31	20.97	22.89
Total	32878	37451	39239	40168	45216	60728	58378	62842
Outward								
Break bulk	1632	1835	1731	1692	1791	2362	1965	2088
% share	20.11	18.25	14.95	13.50	11.93	14.97	10.61	9.89
Growth Rate (%)		12.43	–5.64	–2.23	5.84	31.87	–16.81	6.27
Dry bulk	235	207	177	191	239	363	596	484
% share	2.89	2.06	1.53	1.53	1.59	2.30	3.22	2.29
Growth Rate (%)		–11.70	–14.69	8.06	25.22	51.52	64.27	–18.78
Liquid bulk	389	867	1634	1498	2142	639	1083	1686
% share	4.79	8.63	14.12	11.95	14.27	4.05	5.85	7.89
Growth rate		123.14	88.45	–8.35	43.01	–70.18	69.55	55.64
Containerized cargo	5859	7143	8036	9154	10842	12410	14870	16855
% share	72.21	71.06	69.41	73.02	72.21	78.68	80.32	79.84
Growth rate		21.92	12.49	13.91	18.44	14.46	19.83	13.41
Total	8114	10053	11578	12536	15015	15773	18514	21123
Inward + Outward								
Break bulk	4770	5034	4653	4638	5741	6545	6618	6701
% share	11.64	10.60	9.16	8.80	9.53	9.84	8.61	7.98
Growth Rate (%)		5.53	–7.58	–0.32	23.79	14.01	1.11	1.26
Dry bulk	15652	17743	17470	17105	17088	17725	19181	17518
% share	38.18	37.35	34.38	32.45	28.37	26.65	24.95	20.86
Growth Rate (%)		13.36	-1.54	-2.09	-0.10	3.73	8.21	-8.67
Liquid bulk	6236	7425	9901	9388	11091	11521	14084	15674
% share	15.21	15.83	19.48	17.81	18.41	17.32	18.32	18.67
Growth rate		19.07	33.34	–5.18	18.14	3.88	22.25	11.29
Containerized cargo	14334	17301	18793	21572	26311	30711	37009	44071
% share	34.97	36.42	36.98	40.93	43.68	48.18	48.13	52.49
Growth rate		20.70	8.62	14.79	21.97	16.72	20.51	19.08
Total	40992	47504	50817	52703	60231	66502	76891	83964

Sources:
Hong Kong Port Development Board, *Port Cargo Forecasts*, February 1996, Table C4.
New definitions have been adopted for ocean and river cargoes since 1992.

Table 2.3

Transshipment Ocean Traffic by Cargo Type 1987–1994 ('000 tonnes)

Cargo Type	1987	1988	1989	1990	1991	1992#	1993	1994
Inward								
Break bulk	2164	2089	1807	1302	921	727	483	378
% share	35.68	30.69	27.58	21.43	11.99	8.87	4.91	2.73
Growth Rate (%)		-3.44	-13.49	-27.97	-29.27	-21.07	-33.48	-21.91
Dry bulk	196	267	295	148	188	342	627	763
% share	3.22	3.92	4.50	2.44	2.45	4.17	6.37	5.52
Growth Rate (%)		36.50	10.62	-49.83	27.18	81.62	83.28	21.78
Liquid bulk	12	15	19	12	14	6	61	17
% share	0.19	0.22	0.29	0.20	0.18	0.07	0.62	0.12
Growth Rate (%)		27.75	30.94	-37.54	16.64	- 59.39	961.57	-71.86
Containerized cargo	3693	4436	4431	4612	6560	7120	8678	12673
% share	60.91	65.17	67.62	75.93	85.38	86.89	88.11	91.63
Growth Rate (%)		20.10	-0.10	4.07	42.23	8.54	21.88	46.04
Total	6064	6807	6553	6074	7683	8194	9849	13831
Outward								
Break bulk	1212	979	726	532	411	355	178	143
% share	18.64	13.97	9.96	7.36	4.82	4.06	1.90	1.09
Growth Rate (%)		-19.19	-25.89	-26.69	-22.68	-13.67	-49.87	-19.77
Dry bulk	83	83	81	63	69	45	102	133
% share	1.27	1.18	1.11	0.87	0.81	0.52	1.09	1.01
Growth Rate (%)		0.17	-2.51	-21.91	9.20	-34.25	124.84	30.67
Liquid bulk	9	7	8	9	11	9	8	32
% share	0.14	0.10	0.11	0.13	0.13	0.11	0.09	0.25
Growth Rate (%)		-18.28	7.94	16.98	24.68	-17.73	-14.17	304.50
Containerized cargo	5198	5941	6470	6626	8040	8341	9072	12843
% share	79.95	84.75	88.82	91.64	94.24	95.32	96.92	97.65
Growth Rate (%)		14.29	8.92	2.41	21.33	3.74	8.76	41.58
Total	6501	7010	7285	7231	8532	8750	9360	13152
Inward + Outward								
Break bulk	3375	3068	2533	1834	1332	1082	661	520
% share	26.86	22.21	18.30	13.78	8.22	6.39	3.44	1.93
Growth Rate (%)		-9.09	-17.44	-27.60	-27.36	-18.78	-38.86	-21.33
Dry bulk	278	350	376	211	257	387	729	897
% share	2.22	2.53	2.72	1.59	1.59	2.29	3.80	3.32
Growth Rate (%)		25.69	7.50	-43.82	21.80	50.56	88.14	23.02
Liquid bulk	20	22	27	21	25	15	69	50
% share	0.16	0.16	0.20	0.16	0.16	0.09	0.36	0.18
Growth Rate (%)		7.82	23.39	-21.89	20.10	-40.79	356.31	-27.95
Containerized cargo	8891	10377	10902	11238	14599	15460	17749	25516
% share	70.76	75.10	78.78	84.47	90.04	91.24	92.40	94.56
Growth Rate (%)		16.71	5.08	3.09	29.91	5.90	14.81	43.76
Total	12565	13817	13838	13305	16214	16946	19208	26982

Sources:
Hong Kong Port Development Board, *Port Cargo Forecasts*, February 1996, Table C11.
New definitions have been adopted for ocean and river cargoes since 1992.

Table 2.4
River Traffic by Cargo Type by Shipment Type in 1992–1995
('000 tonnes)

Inward	Direct Shipment				Transhipment			
Cargo Type	1992	1993	1994	1995	1992	1993	1994	1995
Liquid bulk	1	178	232	169	0	3	37	2
% share	0.01	1.66	1.72	1.42	0.00	0.25	1.38	0.08
Dry bulk	7135	7044	10224	8178	0	21	46	20
% share	66.15	65.92	75.83	68.87	0.00	1.88	1.71	0.70
Break bulk	2957	2374	2018	1963	202	100	174	141
% share	27.41	22.22	14.96	16.54	24.07	9.15	6.48	4.96
Containerized	693	1090	1009	1564	638	974	2432	2684
% share	6.42	10.20	7.48	13.17	75.93	88.72	90.44	94.25
Total	10787	10685	13483	11875	841	1098	2689	2848
% share	100	100	100	100	100	100	100	100

Outward	Direct Shipment				Transhipment			
Cargo Type	1992	1993	1994	1995	1992	1993	1994	1995
Liquid bulk	2470	3434	3154	3832	6	2	129	207
% share	38.16	41.06	33.12	35.57	0.48	0.09	2.95	6.40
Dry bulk	209	460	267	265	82	6	594	324
% share	3.24	5.50	2.80	2.46	6.64	0.33	13.56	10.01
Break bulk	2551	3328	4422	4115	313	733	1420	416
% share	39.41	39.79	46.43	38.20	25.42	38.76	32.40	12.84
Containerized	1243	1142	1680	2559	831	1150	2239	2291
% share	19.19	13.65	17.64	23.76	67.45	60.82	51.09	70.75
Total	6473	8365	9523	10771	1233	1891	4383	3238
% share	100	100	100	100	100	100	100	100

Inward + Outward Cargo	Direct Shipment				Transhipment			
Cargo Type	1992	1993	1994	1995	1992	1993	1994	1995
Liquid bulk	2472	3612	3387	4001	6	5	166	210
% share	14.32	18.96	14.72	17.67	0.29	0.15	2.35	3.45
Dry bulk	7345	7504	10491	8443	82	27	640	344
% share	42.55	39.39	45.60	37.28	3.95	0.90	9.05	5.66
Break bulk	5508	5702	6439	6079	516	833	1594	557
% share	31.91	29.93	27.99	26.84	24.87	27.88	22.54	9.15
Containerized	1935	2231	2689	4124	1470	2124	4671	4975
% share	11.21	11.71	11.69	18.21	70.89	71.07	66.05	81.74
Total	17260	19049	23006	22646	2073	2989	7072	6086
% share	100	100	100	100	100	100	100	100

Source: Hong Kong Port Development Board, *Port Cargo Forecasts*, February 1996, Table D3.
1995 data were obtained from Port Development Board's study on River Trade Cargo Handling Activities

Table 2.5

Inward Direct Ocean Traffic by Geographic Zone 1983–1995

('000 tonnes)

Zone of loading	1983	84	85	86	87	88	89	90	91	92#	93	94	95†
Australasia/Oceania	1442	1957	2304	3014	3230	2918	3507	3768	4278	4172	4510	3750	3789
% share	6.60	8.32	8.99	10.00	9.82	7.79	8.94	9.38	9.46	8.22	7.73	5.97	5.31
N. America	1966	1906	2328	2302	2183	2132	2184	2555	3728	3946	3818	4640	5565
% share	8.99	8.11	9.08	7.63	6.64	5.69	5.57	6.36	8.25	7.78	6.54	7.38	7.80
Europe	1127	1566	1727	1829	2107	2561	2178	2294	2652	3038	5366	6653	6374
% share	5.16	6.66	6.74	6.07	6.41	6.84	5.55	5.71	5.87	5.99	9.19	10.59	8.94
Middle East, Africa and Indian Sub-continent*	1852	1966	2980	3578	3613	5941	5954	5133	4289	4609	5269	3803	4456
% share	8.47	8.36	11.62	11.87	10.99	15.86	15.17	12.78	9.48	9.08	9.03	6.05	6.25
ASEAN (Singapore, Malaysia, Thailand, Indonesia, Brunei and Philippines)	4785	5137	5256	6905	7108	8327	9099	9874	10920	12991	14405	17437	21056
% share	21.89	21.84	20.50	22.90	21.62	22.23	23.19	24.58	24.15	25.60	24.68	27.75	29.53
China	3312	3056	2544	3636	4415	4107	4308	5008	5166	5207	5308	5039	5330
% share	15.15	13.00	9.92	12.06	13.43	10.97	10.98	12.47	11.43	10.26	9.09	8.02	7.47
Taiwan	1735	1992	2484	2600	2957	3309	2988	3127	4249	4782	5954	6168	7284
% share	7.94	8.47	9.69	8.63	8.99	8.84	7.62	7.78	9.40	9.43	10.20	9.81	10.21
Japan	4498	4799	4569	4149	4132	4748	5654	5744	6512	7381	8248	9101	9895
% share	20.58	20.41	17.83	13.76	12.57	12.68	14.41	14.30	14.40	14.55	14.13	14.48	13.88
S. Korea	606	628	963	1545	2122	1889	1786	1440	2508	3763	4692	4795	5618
% share	2.77	2.67	3.76	5.12	6.46	5.04	4.55	3.58	5.55	7.42	8.04	7.63	7.88
Others	535	512	478	591	1012	1519	1580	1227	915	848	807	1456	1946
% share	2.45	2.18	1.87	1.96	3.08	4.06	4.03	3.05	2.02	1.67	1.38	2.32	2.73
All zones	21859	23520	25633	30149	32878	37451	39239	40168	45216	50737	58378	62842	71313

Sources:
Hong Kong Port Development Board, *Port Cargo Forecasts*, December 1993; February 1996, Table C5.
Hong Kong Census and Statistics Department, *Hong Kong Shipping Statistics*, October – December 1995, Table C1.

\# New definitions have been adopted for ocean and river cargoes since 1992

* Indian Sub-continent includes India, Pakistan, Sri Lanka, Bangladesh and Burma.

† Figures for 1995 were obtained by grouping shipping statistics compiled by Census and Statistics October – December 1995. However, performing the same exercise" for 1994 failed to yield the figures as reported in the Port Cargo Forecasts.

Table 2.6

Outward Direct Ocean Traffic by Geographic Zone 1983–1995 ('000 tones)

Zone of Discharge	1983	1984	1985	1986	1987	1988	1989	1990	1991	1992#	1993	1994	1995†
Australasia/Oceania	171	211	181	210	283	314	377	339	384	394	425	501	481
% share	3.73	3.95	3.18	2.98	3.49	3.13	3.26	2.71	2.56	2.50	2.30	2.37	2.09
N. America	1015	1216	1197	1471	1771	2031	2526	2678	3251	4049	4692	5531	5879
% share	22.13	22.73	20.97	20.94	21.82	20.21	21.82	21.36	21.65	25.67	25.34	26.19	25.53
Europe	491	520	546	702	980	1196	1304	1775	2215	2519	3064	3360	3568
% share	10.71	9.72	9.57	9.99	12.08	11.90	11.27	14.16	14.75	15.97	16.55	15.91	15.49
Middle East, Africa and Indian Sub-continent*	440	362	346	535	582	672	645	763	837	1018	1260	1213	1036
% share	9.58	6.77	6.06	7.61	7.17	6.68	5.57	6.08	5.57	6.45	6.81	5.74	4.50
ASEAN (Singapore, Malaysia, Thailand, Indonesia, Brunei and Philippines)	970	977	828	1202	1398	1621	1732	1905	1771	1708	1949	2214	2223
% share	21.14	18.27	14.51	17.11	17.23	16.13	14.96	15.20	11.80	10.83	10.53	10.48	9.65
China	154	595	1090	1047	711	1324	1982	1697	2478	2248	3183	3953	4924
% share	3.36	11.13	19.10	14.91	8.76	13.17	17.12	13.54	16.50	14.25	17.19	18.71	21.38
Taiwan	700	715	734	866	911	1241	1400	1669	1962	1983	2029	2050	2034
% share	15.25	13.38	12.85	12.33	11.22	12.35	12.09	13.31	13.07	12.57	10.96	9.71	8.83
Japan	267	303	325	363	576	648	616	675	829	772	814	1034	1227
% share	5.82	5.67	5.70	5.16	7.09	6.44	5.32	5.39	5.52	4.89	4.40	4.89	5.33
S. Korea	169	188	169	270	512	528	486	457	444	329	338	291	272
% share	3.69	3.51	2.96	3.84	6.32	5.25	4.20	3.65	2.96	2.09	1.83	1.38	1.18
Others	211	260	291	360	391	477	510	577	844	754	758	975	1387
% share	4.59	4.87	5.09	5.13	4.82	4.74	4.41	4.60	5.62	4.78	4.09	4.62	6.02
All zones	4589	5347	5707	7026	8114	10053	11578	12535	15015	15773	18514	21123	23031

Sources:
Hong Kong Port Development Board, *Port Cargo Forecasts*, December 1993; February 1996, Table C6.
Hong Kong Census and Statistics Department, Hong Kong Shipping Statistics, October–December 1995, Table C2.

\# New definitions have been adopted for ocean and river cargoes since 1992.
* Indian Sub-continent includes India, Pakistan, Sri Lanka, Bangladesh and Burma.
† Figures for 1995 were obtained by grouping shipping statistics compiled by Census and Statistics October - December 1995. However, performing the same exercise for 1994 failed to yield the figures as reported in the Port Cargo Forecasts.

Table 2.7
Inward Containerized Direct Ocean Traffic by Geographic Zone 1987–1995 ('000 tonnes)

Zone of loading	1987	1988	1989	1990	1991	1992#	1993	1994	1995†
Australasia/Oceania	356	385	482	405	543	607	768	730	989
% share	4.20	3.79	4.48	3.27	3.51	3.32	3.47	2.68	3.17
N. America	1482	1773	1807	2115	2698	3164	3133	4215	5317
% share	17.49	17.46	16.80	17.03	17.44	17.29	14.15	15.49	17.05
Europe	1464	1796	1665	1736	1993	2296	3180	4289	4227
% share	17.28	17.68	15.48	13.98	12.89	12.54	14.36	15.77	13.56
Middle East, Africa and Indian Sub-continent*	273	322	439	466	532	637	704	846	776
% share	3.22	3.17	4.09	3.75	3.44	3.48	3.18	3.11	2.49
ASEAN (Singapore, Malaysia, Thailand, Indonesia, Brunei and Philippines)	1232	1452	1503	1671	1949	2355	2965	3491	4358
% share	14.54	14.30	13.98	13.46	12.60	12.87	13.39	12.83	13.98
China	173	269	352	461	592	842	1345	1823	2229
% share	2.04	2.65	3.27	3.71	3.83	4.60	6.07	6.70	7.15
Taiwan	1190	1483	1899	2518	3518	3957	4921	5411	6282
% share	14.04	14.60	17.65	20.28	22.74	21.62	22.23	19.89	20.15
Japan	1442	1560	1527	1894	2180	2587	2863	3423	3507
% share	17.02	15.36	14.20	15.25	14.09	14.14	12.93	12.58	11.25
S. Korea	729	955	931	989	1297	1642	2096	2495	2678
% share	8.61	9.40	8.65	7.97	8.38	8.97	9.47	9.17	8.59
Others	133	164	151	162	167	214	166	484	814
% share	1.56	1.61	1.41	1.31	1.08	1.17	0.75	1.78	2.61
All zones	8475	10158	10758	12418	15469	18301	22139	27206	31177

Sources:

1. Hong Kong Port Development Board, *Port Cargo Forecasts*, February 1996, Table C25.
2. Hong Kong Census and Statistics Department, Hong Kong Shipping Statistics, October – December 1995, Table C3.

\# New definitions have been adopted for ocean and river cargoes since 1992.

* Indian Sub-continent includes India, Pakistan, Sri Lanka, Bangladesh and Burma.

† Figures for 1995 were obtained by grouping shipping statistics compiled by Census and Statistics October – December 1995. However, performing the same exercise for 1994 failed to yield the figures as reported in the Port Cargo Forecasts.

Table 2.8
Outward Containerized Direct Ocean Traffic by Geographic Zone 1987–1995 ('000 tonnes)

Zone of Discharge	1987	1988	1989	1990	1991	1992#	1993	1994	1995†
Australasia/Oceania	278	312	368	334	376	393	425	500	479
% share	4.74	4.36	4.58	3.65	3.46	3.16	2.86	2.97	2.66
N. America	1770	2029	2525	2677	3250	4044	4686	5531	5874
% share	30.21	28.40	31.42	29.24	29.97	32.58	31.52	32.80	32.64
Europe	973	1195	1303	1772	2212	2517	3061	3360	3563
% share	16.61	16.73	16.22	19.35	20.41	20.28	20.58	19.92	19.80
Middle East, Africa and Indian Sub-continent*	511	598	611	737	816	1003	1251	1211	1024
% share	8.73	8.37	7.61	8.06	7.52	8.09	8.41	7.18	5.69
ASEAN (Singapore, Malaysia, Thailand, Indonesia, Brunei and Philippines)	767	981	1120	1351	1352	1405	1720	2068	1978
% share	13.10	13.73	13.94	14.75	12.47	11.32	11.56	12.26	10.99
China	154	314	275	225	305	517	939	1035	1391
% share	2.62	4.39	3.43	2.46	2.82	4.17	6.31	6.14	7.73
Taiwan	394	516	594	742	930	979	992	1052	1160
% share	6.72	7.22	7.39	8.11	8.58	7.89	6.67	6.24	6.44
Japan	490	598	570	606	770	751	789	922	1082
% share	8.36	8.38	7.09	6.62	7.10	6.05	5.31	5.47	6.01
S. Korea	360	414	393	384	381	278	293	257	244
% share	6.14	5.80	4.89	4.20	3.51	2.24	1.97	1.52	1.36
Others	162	186	277	326	451	524	714	928	1204
% share	2.77	2.61	3.45	3.56	4.16	4.22	4.80	5.51	6.69
All zones	5859	7143	8036	9154	10842	12410	14870	16865	17999

Sources:
1. Hong Kong Port Development Board, *Port Cargo Forecasts*, February 1996, Table C26.
2. Hong Kong Census and Statistics Department, Hong Kong Shipping Statistics, October – December 1995, Table C4.

\# New definitions have been adopted for ocean and river cargoes since 1992

* Indian Sub-continent includes India, Pakistan, Sri Lanka, Bangladesh and Burma.

† Figures for 1995 were obtained by grouping shipping statistics compiled by Census and Statistics October - December 1995. However, performing the same exercise for 1994 failed to yield the figures as reported in the Port Cargo Forecasts.

Table 2.9
Inward Transshipment Ocean Traffic by Geographic Zone 1983–1995
('000 tonnes)

Zone of loading	1983	84	85	86	87	88	89	90	91	92#	93	94	95†
Australasia/ Oceania	27	73	185	218	166	207	131	120	170	196	289	397	519
% share	1.17	2.49	4.61	4.40	2.74	3.05	1.99	1.97	2.21	2.39	2.93	2.87	3.30
N. America	592	825	1248	939	1463	1443	1365	1008	1375	1133	1088	1961	2753
% share	25.96	28.14	31.02	18.96	24.13	21.20	20.83	16.60	17.90	13.82	11.05	14.18	17.50
Europe	236	373	494	644	788	1027	940	832	1256	1038	1203	1850	1407
% share	10.36	12.74	12.27	13.01	13.00	15.09	14.34	13.69	16.35	12.66	12.22	13.38	8.94
Middle East, Africa and Indian Sub-continent*	58	81	112	187	413	332	303	280	202	204	182	297	224
% share	2.56	2.75	2.78	3.78	6.80	4.88	4.63	4.61	2.63	2.49	1.85	2.15	1.42
ASEAN (Singapore, Malaysia, Thailand, Indonesia, Brunei and Philippines)	286	355	457	661	609	679	636	563	727	922	899	1470	1796
% share	12.55	12.09	11.35	13.36	10.04	9.97	9.71	9.27	9.46	11.25	9.12	10.63	11.41
China	784	883	1046	1538	1810	1946	2123	2195	2379	2616	3363	4573	5252
% share	34.39	30.10	25.99	31.05	29.84	28.59	32.39	36.13	30.97	31.93	34.14	33.07	33.38
Taiwan	103	164	213	268	240	214	276	322	660	1081	1403	1466	1918
% share	4.50	5.58	5.29	5.42	3.96	3.15	4.21	5.29	8.59	13.19	14.25	10.60	12.19
Japan	96	109	115	238	213	233	217	265	588	661	911	1037	995
% share	4.20	3.72	2.87	4.81	3.52	3.42	3.31	4.36	7.66	8.07	9.25	7.50	6.32
S. Korea	28	24	46	124	44	90	135	143	185	214	346	499	441
% share	1.22	0.83	1.13	2.50	0.72	1.33	2.06	2.35	2.41	2.61	3.51	3.61	2.80
Others	70	45	109	134	318	635	427	348	139	130	165	280	430
% share	3.09	1.55	2.70	2.71	5.25	9.33	6.51	5.72	1.81	1.58	1.68	2.03	2.73
All zones	2279	2932	4024	4952	6064	6807	6553	6074	7683	8194	9849	13831	15735

Sources:
1. Hong Kong Port Development Board, *Port Cargo Forecasts*, December 1993; February 1996, Table C12.
2. Hong Kong Census and Statistics Department, Hong Kong Shipping Statistics, October–December 1995, Table C1.

\# New definitions have been adopted for ocean and river cargoes since 1992.
* Indian Sub-continent includes India, Pakistan, Sri Lanka, Bangladesh and Burma.
† Figures for 1995 were obtained by grouping shipping statistics compiled by Census and Statistics October – December 1995. However, performing the same exercise for 1994 failed to yield the figures as reported in the Port Cargo Forecasts.

Table 2.10

Outward Transshipment Ocean Traffic by Geographic Zone 1983–1995

('000 tonnes)

Zone of Discharge	1983	84	85	86	87	88	89	90	91	92#	93	94	95†
Australasia/ Oceania	124	132	148	191	230	233	245	220	225	249	225	299	541
% share	4.36	3.78	3.42	3.57	3.54	3.32	3.36	3.04	2.63	2.84	2.40	2.27	3.16
N. America	478	643	670	785	885	917	1042	1083	1201	1408	1315	1840	2166
% share	16.83	18.40	15.49	14.69	13.62	13.08	14.31	14.97	14.08	16.09	14.05	13.99	12.67
Europe	488	583	670	1040	1312	1268	1414	1585	1710	1663	1879	2614	3427
% share	17.18	16.69	15.49	19.46	20.19	18.09	19.41	21.93	20.04	19.01	20.07	19.87	20.05
Middle East, Africa and Indian Sub-continent*	451	411	481	658	735	722	713	736	684	731	1013	978	1074
% share	15.87	11.76	11.12	12.32	11.31	10.31	9.79	10.18	8.01	8.35	10.82	7.44	6.28
ASEAN (Singapore, Malaysia, Thailand, Indonesia, Brunei and Philippines)	835	864	702	998	1134	1182	1462	1527	1601	1555	1457	2197	2788
% share	29.39	24.73	16.23	18.69	17.45	16.87	20.07	21.11	18.76	17.77	15.57	16.71	16.31
China	102	300	1043	829	1270	1548	1356	924	1625	1754	2048	3147	4190
% share	3.59	8.59	24.12	15.52	19.54	22.08	18.62	12.78	19.05	20.05	21.88	23.93	24.51
Taiwan	154	256	236	267	383	467	495	469	542	430	421	789	818
% share	5.42	7.33	5.45	5.00	5.90	6.66	6.80	6.49	6.36	4.92	4.50	6.00	4.78
Japan	109	179	222	341	340	403	317	375	479	488	477	587	882
% share	3.84	5.12	5.12	6.38	5.23	5.75	4.35	5.19	5.61	5.58	5.10	4.47	5.16
S. Korea	33	62	90	126	97	138	87	127	231	167	179	237	335
% share	1.16	1.77	2.09	2.37	1.49	1.97	1.20	1.76	2.70	1.91	1.91	1.80	1.96
Others	65	64	64	106	113	131	151	185	235	304	345	463	875
% share	2.29	1.83	1.48	1.99	1.74	1.88	2.07	2.56	2.75	3.47	3.68	3.52	5.12
Total	2841	3494	4326	5342	6501	7010	7285	7231	8532	8750	9360	13152	17096

Sources:

1. Hong Kong Port Development Board, *Port Cargo Forecasts*, December 1993; February 1996, Table C13.
2. Hong Kong Census and Statistics Department, Hong Kong Shipping Statistics, October – December 1995, Table C2.

1992 and onwards data are in new definitions.

* Indian Sub-continent includes India, Pakistan, Sri Lanka, Bangladesh and Burma.

† Figures for 1995 were obtained by grouping shipping statistics compiled by Census and Statistics October – December 1995. However, performing the same exercise for 1994 failed to yield the figures as reported in the Port Cargo Forecasts.

Table 2.11
Inward Containerized Transshipment Ocean Traffic by Geographic Zone 1987–1995 ('000 tonnes)

Zone of loading	1987	1988	1989	1990	1991	1992#	1993	1994	1995†
Australasia/Oceania	156	172	114	115	154	186	258	392	513
% share	4.22	3.88	2.57	2.49	2.35	2.61	2.97	3.09	3.38
N.America	1019	1237	1116	1000	1303	1110	1047	1937	2731
% share	27.57	27.88	25.18	21.69	19.87	15.59	12.06	15.29	17.98
Europe	634	879	888	801	1223	967	967	1524	1352
% share	17.15	19.81	20.04	17.37	18.65	13.58	11.14	12.03	8.90
Middle East, Africa & Indian Sub-continent*	136	207	173	163	196	201	167	270	215
% share	3.68	4.67	3.90	3.54	2.99	2.82	1.92	2.13	1.42
ASEAN (Singapore, Malaysia Thailand, Indonesia & Brunei, Philippines)	572	547	458	476	661	860	838	1394	1743
% share	15.48	12.33	10.33	10.32	10.08	12.08	9.66	11.00	11.48
China	626	782	1034	1287	1612	2012	3029	4385	5165
% share	16.94	17.62	23.33	27.91	24.58	28.26	34.90	34.60	34.00
Taiwan	235	211	274	314	658	1071	1389	1431	1886
% share	6.36	4.76	6.18	6.81	10.03	15.04	16.01	11.29	12.42
Japan	183	208	174	230	484	512	697	814	843
% share	4.95	4.69	3.93	4.99	7.38	7.19	8.03	6.42	5.55
S. Korea	41	82	99	113	158	140	198	313	349
% share	1.11	1.85	2.23	2.45	2.41	1.97	2.28	2.47	2.30
Others	94	112	102	112	110	60	88	212	392
% share	2.54	2.52	2.30	2.43	1.68	0.84	1.01	1.67	2.58
All zones	3696	4437	4432	4611	6559	7119	8678	12672	15189

Sources:
1. Hong Kong Port Development Board, *Port Cargo Forecasts*, February 1996, Table C27.
2. Hong Kong Census and Statistics Department, Hong Kong Shipping Statistics, October – December 1995, Table C3.

\# New definitions have been adopted for ocean and river cargoes since 1992.

* Indian Sub-continent includes India, Pakistan, Sri Lanka, Bangladesh and Burma.

† Figures for 1995 were obtained by grouping shipping statistics compiled by Census and Statistics October – December 1995. However, performing the same exercise for 1994 failed to yield the figures as reported in the Port Cargo Forecasts.

Table 2.12

Outward Containerized Transshipment Ocean Traffic by Geographic Zone 1987–1995 ('000 tonnes)

Zone of loading	1987	1988	1989	1990	1991	1992#	1993	1994	1995†
Australasia/Oceania	229	232	244	215	220	248	223	298	540
% share	4.41	3.91	3.77	3.24	2.74	2.97	2.46	2.32	3.19
N.America	884	916	1042	1083	1201	1408	1315	1840	2165
% share	17.01	15.42	16.11	16.34	14.94	16.88	14.49	14.33	12.78
Europe	1276	1253	1402	1580	1703	1660	1878	2611	3424
% share	24.56	21.09	21.67	23.84	21.18	19.90	20.70	20.33	20.22
Middle East, Africa & Indian Sub-continent*	484	516	556	598	603	672	990	968	1074
% share	9.31	8.69	8.59	9.02	7.50	8.06	10.91	7.54	6.34
ASEAN (Singapore, Malaysia, Thailand, Indonesia & Brunei, Philippines)	845	933	1222	1292	1351	1357	1366	2146	2746
% share	16.26	15.70	18.89	19.50	16.80	16.27	15.05	16.71	16.21
China	608	991	987	734	1503	1671	1922	2964	4095
% share	11.70	16.68	15.26	11.08	18.69	20.03	21.18	23.08	24.18
Taiwan	363	457	484	465	537	384	390	763	807
% share	6.99	7.69	7.48	7.02	6.68	4.60	4.30	5.94	4.77
Japan	337	399	317	375	478	488	477	558	881
% share	6.49	6.72	4.90	5.66	5.95	5.85	5.26	4.35	5.20
S. Korea	94	138	87	123	229	165	178	237	332
% share	1.81	2.32	1.34	1.86	2.85	1.98	1.96	1.85	1.96
Others	76	106	129	162	215	289	335	457	871
% share	1.46	1.78	1.99	2.44	2.67	3.46	3.69	3.56	5.14
All zones	5196	5941	6470	6627	8040	8342	9074	12842	16935

Sources:

1. Hong Kong Port Development Board, *Port Cargo Forecasts*, February 1996, Table C28.
2. Hong Kong Census and Statistics Department, Hong Kong Shipping Statistics, October – December 1995, Table C4.

\# New definitions have been adopted for ocean and river cargoes since 1992.

* Indian Sub-continent includes India, Pakistan, Sri Lanka, Bangladesh and Burma.

† Figures for 1995 were obtained by grouping shipping statistics compiled by Census and Statistics October - December 1995. However, performing the same exercise for 1994 failed to yield the figures as reported in the Port Cargo Forecasts.

Table 2.13

Hong Kong – China Port Cargo Movement by Modes of Transport 1983–1995

('000 tonnes)

Year	1983	84	85	86	87	88	89	90	91	92#	93	94	95
Inward Cargo													
Ocean	4096	3939	3590	5174	6224	6053	6431	7202	7545	7823	8671	9612	10582
Ocean Growth Rate (%)		–3.83	–8.86	44.13	20.30	–2.76	6.25	11.99	4.77	3.68	10.83	10.86	10.09
River	3525	3564	4589	5713	5786	5697	5140	5770	6505	11361	11410	15801	14404
River Growth Rate (%)		1.11	28.77	24.48	1.29	–1.54	-9.79	12.27	12.74	74.65	0.43	38.48	–8.84
Ocean and River	7621	7503	8179	10887	12011	11750	11571	12972	14051	19185	20081	25413	24986
Ocean and River Growth Rate (%)		–1.55	9.01	33.10	10.32	–2.17	–1.52	12.11	8.31	36.54	4.67	26.56	–1.68
Ocean/(Ocean + River) (%)	53.75	52.50	43.89	47.53	51.82	51.51	55.58	55.52	53.70	40.78	43.18	37.82	42.35
All Modes of Transport †	9753	10177	11187	15014	17066	17200	17597	19568	21469	27319	28918	34245	NA
(Ocean + River) / All Modes of Transport (%)	78.14	73.73	73.11	72.51	70.38	68.32	65.76	66.29	65.45	70.23	69.44	74.21	NA
Outward Cargo													
Ocean	257	895	2133	1877	1981	2872	3338	2622	4103	4002	5231	7099	9114
Ocean Growth Rate (%)		248.25	138.36	–12.03	5.55	44.98	16.23	–21.46	56.50	–2.45	30.70	35.71	28.38
River	816	1226	2026	1910	2581	3200	2658	2637	3839	6776	8964	12235	12474
River Growth Rate (%)		50.25	65.22	–5.70	35.14	23.97	–16.92	–0.82	45.62	76.50	32.28	36.49	1.95
Ocean and River	1073	2121	4159	3787	4562	6072	5996	5258	7942	10779	14195	19334	21588
Ocean and River Growth Rate (%)		97.67	96.09	–8.95	20.48	33.09	–1.24	–12.31	51.04	35.71	31.70	36.20	11.66
Ocean/(Ocean + River) (%)	23.95	42.20	51.29	49.56	43.42	47.30	55.67	49.86	51.66	37.13	36.85	36.72	42.22

Table 2.13 (continued)

Year	1983	84	85	86	87	88	89	90	91	92#	93	94	95
All Modes of Transport	1958	3293	5620	5866	7387	9784	10318	10134	13789	16902	20273	25317	NA
(Ocean + River) / All Modes of Transport (%)	54.80	64.41	74.01	64.56	61.75	62.06	58.11	51.89	57.60	63.77	70.02	76.37	NA
Inward + Outward Cargo													
Ocean	4352	4834	5723	7051	8205	8925	9769	9824	11648	11826	13902	16711	19696
Ocean Growth Rate (%)		11.08	18.39	23.19	16.38	8.77	9.46	0.56	18.57	1.52	17.56	20.21	17.86
River	4341	4791	6615	7623	8368	8897	7798	8407	10344	18138	20374	28036	26878
River Growth Rate (%)		10.37	38.07	15.24	9.77	6.33	−12.35	7.81	23.05	75.34	12.33	37.61	−4.13
Ocean and River	8693	9625	12338	14674	16573	17822	17567	18231	21993	29964	34276	44747	46574
Ocean and River Growth Rate (%)		10.72	28.19	18.93	12.94	7.54	−1.43	3.78	20.64	36.24	14.39	30.55	4.08
Ocean/(Ocean + River) (%)	50.06	50.22	46.39	48.05	49.51	50.08	55.61	53.89	52.96	39.47	40.56	37.35	42.29
All Modes of Transport	11711	13470	16807	20879	24453	26984	27915	29702	35258	44220	49192	59563	NA
(Ocean + River) / All Modes of Transport (%)	74.23	71.46	73.41	70.28	67.77	66.05	62.93	61.38	62.38	67.76	69.68	75.13	NA

Sources:
1. Hong Kong Port Development Board, *Port Cargo Forecasts*, December 1993; February 1996, Table A2.
2. 1995 ocean data were obtined from Hong Kong Census and Statistics Department, Hong Kong Shipping Statistics, October–December 1995, Table C1, C2.
3. 1995 river data were obtained from Port Development Board's study on River Trade Cargo Handling Activities.

Note:
Inward cargo figures exclude gravel & crushed stone imported by conveyor transport system across the border.
New definitions have been adopted for ocean and river cargoes since 1992.
† The modes of transport include ocean, river, rail, road, and air.

Table 2.14
Percentage Share of Freight Movement from China by Ocean and River 1983–1995

Year	1983	84	85	86	87	88	89	90	91	92#	93	94	95
Inward Cargo													
Ocean and River	26.85	24.43	23.42	26.29	26.64	23.38	22.57	24.82	23.57	27.19	25.10	27.37	24.55
Outward Cargo													
Ocean and River	12.47	20.15	32.92	25.46	25.53	28.75	26.75	22.83	28.39	33.44	37.23	40.13	39.88
Inward + Outward Cargo													
Ocean and River	23.50	23.34	25.95	26.07	26.32	24.96	23.84	24.21	25.11	29.15	29.01	31.73	29.87

Sources:
1. Hong Kong Port Development Board, *Port Cargo Forecasts*, February 1996, Table A1, A2.
2. Hong Kong Census and Statistics Department, Hong Kong Shipping Statistics, January – March 1996.

Note:
New definitions have been adopted for ocean and river cargoes since 1992.

Table 2.15
Hong Kong – China Ocean Traffic by Port of Loading/Discharge 1985–1994 ('000 tonnes)

Year	1985	1986	1987	1988	1989	1990	1991	1992#	1993	1994
Inward Cargo										
Ports of Loading										
North China	1571	2581	3371	3140	3236	3638	3700	3622	3572	3237
Middle China	1090	1536	1720	1786	1800	1971	2083	1967	2451	3240
South China	660	681	671	692	736	757	773	1276	1284	1671
Other ports	270	376	462	436	659	837	990	958	1364	1464
Total	3590	5174	6224	6053	6431	7202	7545	7823	8671	9612
Outward Cargo										
Ports of Discharge										
North China	537	369	444	578	456	251	378	391	415	539
Middle China	998	946	788	1018	853	583	854	921	1056	1357
South China	371	206	345	353	538	460	755	1962	3004	3501
Other ports	228	357	403	923	1492	1328	2116	728	756	1702
Total	2133	1877	1981	2872	3338	2622	4103	4002	5231	7099
Inward + Outward cargo										
Ports of Loading/Discharge										
North China	2107	2949	3816	3717	3691	3889	4078	4014	3987	3776
Middle China	2087	2482	2508	2804	2652	2553	2937	2888	3507	4597
South China	1031	887	1017	1045	1274	1216	1528	3238	4288	5173
Other ports	497	733	865	1359	2151	2165	3106	1687	2120	3166
Total	5723	7051	8205	8925	9769	9824	11648	11826	13902	16711

Sources:
1. Hong Kong Port Development Board, *Port Cargo Forecasts*, February 1996, Table C5A, C5B, C5C.

Note:
\# New definitions have been adopted for ocean and river cargoes since 1992.

North China ports: Dalian, Qinhuangdao, Tianjin Xingang
Middle China ports: Nanjing, Qingdao, Shanghai, Wenzhou, Wuhu, Zhangjiagang
South China ports: Fuzhou, Haikou, Shantou, Xiamen, Zhanjiang

Table 2.16
China's Exports and Imports by Region (US$ 100 Million)

	Exports			Imports			Total Trade		
	1992	1993	1994	1992	1993	1994	1992	1993	1994
Region									
North China	73.62	78.85	103.48	88.52	116.47	140.31	162.14	195.32	243.80
% share	8.67	8.59	8.55	10.98	11.20	12.14	9.80	9.98	10.30
Northeast China	96.34	92.32	97.12	59.53	78.36	85.42	155.87	170.68	182.54
% share	11.34	10.06	8.03	7.39	7.54	7.39	9.42	8.72	7.71
East China	256.37	286.67	375.82	237.30	308.60	321.05	493.67	595.27	696.87
% share	30.18	31.25	31.06	29.45	29.68	27.77	29.82	30.42	29.45
Total of South China	387.43	422.29	588.93	387.93	493.08	549.17	775.36	915.37	1138.10
% share	45.61	46.03	48.67	48.14	47.43	47.50	46.84	46.77	48.10
Southwest China	20.66	22.71	26.94	19.48	22.59	37.95	40.14	45.31	64.88
% share	2.43	2.48	2.23	2.42	2.17	3.28	2.43	2.32	2.74
Northwest China	14.97	14.60	17.77	13.10	20.49	22.24	28.07	35.09	40.01
% share	1.76	1.59	1.47	1.63	1.97	1.92	1.70	1.79	1.69
Total	849.40	917.44	1210.06	805.85	1039.59	1156.14	1655.25	1957.03	2366.20

Sources:
China's Customs Statistics Yearbook 1992; 1993; 1994.

Table 2.17

Containerized Cargoes: Hong Kong–China Ocean Traffic by Port of Loading/Discharge 1992–1994 ('000 tonnes)

	Inward			Outward			Total		
	1992	1993	1994	1992	1993	1994	1992	1993	1994
Ports of Loading/Discharge									
North China	620	860	1221	363	407	518	983	1267	1739
% share	21.73	19.66	19.67	16.60	14.22	12.96	19.50	17.51	17.04
Middle China	1321	2115	2797	861	998	1308	2182	3112	4105
% share	46.29	48.35	45.05	39.36	34.87	32.72	43.28	43.02	40.22
South China	601	927	1258	809	1242	1721	1410	2168	2980
% share	21.07	21.19	20.27	36.97	43.40	43.05	27.97	29.97	29.19
Other ports	312	472	932	155	215	451	466	687	1383
% share	10.91	10.80	15.01	7.08	7.51	11.28	9.25	9.50	13.55
Total	2854	4373	6208	2188	2861	3999	5042	7234	10207

Source:
Hong Kong Port Development Board, *Port Cargo Forecasts*, February 1996, Table C39.

North China ports: Dalian, Qinhuangdao, Tianjin Xingang
Middle China ports : Nanjing, Qingdao, Shanghai, Wenzhou, Wuhu, Zhangjiagang
South China ports: Fuzhou, Haikou, Shantou, Xiamen, Zhanjiang

Table 2.18

Total (Inward and Outward) River Traffic by Port of Loading/Discharge by Shipment Type 1992–1995 ('000 tonnes)

Port of Loading/Discharge	1992	1993	1994	1995
Direct Shipment				
Dongguan	467	1170	1268	1788
Foshan	331	535	608	482
Guangzhou	2068	2295	2059	2149
Huizhou	16	3	24	11
Jiangmen/xinhui	840	1304	1117	1345
Nanhai	264	273	406	986
Panyu	226	160	727	1081
Shenzhen	2588	3391	3051	3097
Shunde	305	346	692	574
Zhaoqing/Yunfu	205	165	196	244
Zhongshan	517	1118	1013	1507
Zhuhai	6511	6331	8745	7505
Guangxi	903	520	1972	678
Macau	856	1422	1102	1161
Others	1162	16	28	36
Total	17260	19049	23006	22646
Transhipment				
Dongguan	9	47	27	16
Foshan	145	228	321	300
Guangzhou	981	1377	2784	2857
Huizhou	0	0	28	0
Jiangmen/xinhui	118	107	237	226
Nanhai	3	3	53	144
Panyu	11	58	158	227
Shenzhen	95	419	1252	691
Shunde	52	53	120	166
Zhaoqing/Yunfu	25	15	64	56
Zhongshan	102	200	300	278
Zhuhai	129	206	564	303
Guangxi	29	33	156	74
Macau	340	242	941	693
Others	36	3	66	55
Total	2073	2989	7072	6086
Total				
Dongguan	476	1217	1295	1805
Foshan	476	763	929	782
Guangzhou	3049	3672	4844	5005
Huizhou	16	3	52	11
Jiangmen/xinhui	958	1411	1354	1572
Nanhai	267	276	459	1130
Panyu	237	218	885	1309
Shenzhen	2683	3810	4303	3788
Shunde	357	399	811	740
Zhaoqing/Yunfu	230	180	261	301
Zhongshan	619	1317	1312	1785
Zhuhai	6640	6537	9309	7808
Guangxi	932	553	2127	752
Macau	1195	1664	2043	1854
Others	1198	19	94	91
Total	19333	22038	30079	28732

Sources:
Hong Kong Port Development Board, *Port Cargo Forecasts*, February 1996, Table D4, D5.
1995 data were obtained from Port Development Board's study on River Trade Cargo Handling Activities.

Table 2.19

Hong Kong's Overall Container Throughput 1983–1996 ('000 TEUs)

Year	1983	84	85	86	87	88	89	90	91	92#@	93	94	95	96**
Inward Traffic														
Container terminals	812	902	945	1066	1290	1465	1601	1830	2161	2428	2750	3397	3904	1883
% growth		11.08	4.76	12.84	20.93	13.59	9.30	14.30	18.10	12.36	13.23	23.54	14.95	3.31
Mid-stream & other-terminals*	107	163	209	273	416	522	575	663	861	1284	1503	1595	1612	782
% growth		52.34	28.33	30.48	52.48	25.48	10.07	15.40	29.82	49.07	17.12	6.12	1.07	–2.74
River trade vessels##	NA	NA	NA	41	29	41	46	37	37	220	314	477	698	377
% growth	NA	NA	NA	NA	–27.54	39.91	12.30	–18.91	–0.62	489.78	42.69	52.06	46.41	34.44
Total	NA	NA	NA	1380	1735	2028	2222	2531	3059	3932	4566	5469	6215	3042
% growth					25.75	16.89	9.56	13.89	20.89	28.50	16.15	19.76	13.64	4.63
Outward Traffic														
Container terminals	824	900	950	1123	1324	1537	1716	2001	2353	2651	3047	3881	4352	2058
% growth		9.22	5.54	18.24	17.89	16.12	11.62	16.62	17.59	12.64	14.95	27.38	12.12	0.40
Mid-stream & other-terminals*	95	144	185	238	364	428	492	535	712	1178	1294	1244	1317	670
% growth		51.58	28.44	28.46	53.17	17.52	15.16	8.67	33.05	65.36	9.90	-3.88	5.88	6.58
River trade vessels##	NA	NA	NA	33	34	40	33	33	37	212	297	456	666	363
% growth					2.02	17.92	–17.74	1.05	10.97	471.33	39.93	53.64	46.08	25.94
Total	NA	NA	NA	1394	1722	2005	2242	2570	3102	4040	4638	5581	6335	3090
% growth					23.52	16.45	11.79	14.64	20.72	30.23	14.79	20.34	13.50	4.19

Inward + Outward Traffic														
Container terminals	1636	1802	1895	2189	2614	3002	3317	3831	4514	5079	5797	7278	8256	3941
% growth	8.52	10.15	5.16	15.55	19.37	14.87	10.49	15.50	17.83	12.50	14.13	25.56	13.44	1.77
Mid-stream & other-terminals*	201	307	394	511	780	950	1067	1198	1573	2461	2797	2839	2930	1452
% growth		52.74	28.38	29.53	52.80	21.76	12.36	12.29	31.26	56.44	13.67	1.50	3.18	1.35
River trade vessels##	NA	NA	NA	74	64	81	79	71	74	432	610	933	1364	739
% growth					–14.19	28.10	–2.55	–10.58	4.84	480.57	41.34	52.83	46.25	30.13
Total	NA	NA	NA	2774	3457	4033	4464	5101	6162	7972	9204	11050	12550	6132
% growth					24.63	16.67	10.67	14.27	20.81	29.37	15.46	20.05	13.57	4.41
% Share of Total†														
Container terminals	89.06	85.44	82.78	78.93	75.60	74.43	74.31	75.11	73.26	63.71	62.98	65.87	65.79	64.26
Mid-stream & other-terminals*	10.94	14.56	17.22	18.40	22.56	23.55	23.91	23.50	25.53	30.87	30.39	25.69	23.34	23.68
River trade vessels##	NA	NA	NA	2.67	1.84	2.02	1.78	1.39	1.21	5.42	6.63	8.44	10.87	12.06
Total	NA	NA	NA	100	100	100	100	100	100	100	100	100	100	100

Sources:
1985–994: Hong Kong Port Development Board, Port Cargo Forecasts, December 1993; February 1996, Table E1.
1983–984, 1995, 1996: Marine Department

* Exclude containers handled by terminal vessels contracted to work in stream
Exclude Kwai Chung Terminals
@ A standardised manifest has been introduced from 1 April 92 to collect river cargo and container throughput data
# New definitions have been adopted for ocean and river containers since 1992.
† The share figures for 1983–1985 were the relative shares of terminal and mid-stream only.
** 1996 data include throughput of the first six months, and the growth rate is measured with respect to the same period of 1995.
NA not available

Table 2.20

Percentage Share of Transshipment Containers (in TEU) 1985–1995

	1985	1986	1987	1988	1989	1990	1991	1992#	1993	1994	1995
Kwai Chung Terminals											
Inward	27.99	27.37	26.36	25.70	21.49	20.97	21.05	17.84	15.97	23.74	28.70
Outward	26.44	25.96	25.02	24.09	19.77	18.81	18.53	16.00	13.42	20.53	26.02
Total	27.22	26.65	25.68	24.87	20.60	19.84	19.73	16.88	14.63	22.03	27.29
Mid-Stream & other terminals											
Inward	NA	NA	NA	NA	NA	NA	NA	7.53	7.46	7.13	9.97
Outward	NA	NA	NA	NA	NA	NA	NA	5.94	7.44	5.82	9.38
Total	NA	NA	NA	NA	NA	NA	NA	6.76	7.45	6.56	9.70
River trade vessels											
Inward	NA	NA	NA	NA	NA	NA	NA	22.28	19.00	20.49	19.73
Outward	NA	NA	NA	NA	NA	NA	NA	16.17	16.21	15.52	16.66
Total	NA	NA	NA	NA	NA	NA	NA	19.28	17.64	18.06	18.23
Total											
Inward	NA	NA	NA	NA	NA	NA	NA	14.72	13.38	18.61	22.83
Outward	NA	NA	NA	NA	NA	NA	NA	13.08	11.93	16.85	21.57
Total	NA	NA	NA	NA	NA	NA	NA	13.89	12.65	17.72	22.20

Source: Hong Kong Port Development Board, *Port Cargo Forecasts*, February 1996, Table E6, Marine Department.

Bibliography

1. Chartered International (1996). *The Port of Hong Kong 1996*. London,.

2. Chu, David Y. K. (1994) "Challenges to the Port of Hong Kong Before and After 1997," *Chinese Environment and Development* Vol. 5, No. 3, (1994), pp. 5–23.

3. Council for Economic Planning and Development (1994). *Sea Shipping and Value-Added Operations Center*. Taipei, Taiwan, December.

4. _______ (1995). Developing Taiwan into a Regional Operations Center (in Chinese). Taipei, Taiwan, 5 January.

5. Hong Kong Centre for Economic Research (1992). "Competition in Container Terminals," *HKCER Letters* Vol. 16, September.

6. International Monetary Fund (1996). *World Economic Outlook*. Washington, D.C., May.

7. Kaohsiung Harbor Bureau (1996). *Port of Kaohsiung 1996*. Kaohsiung, Taiwan.

8. Lam, Pun-lee (1996). *The Scheme of Control on Electricity Companies*. Hong Kong: The Chinese University Press.

9. Lau, Lawrence J. (1995). "*Long-Term Economic Growth in the PRC and its Sectoral Implications,*" in *The Economy of the PRC: Analysis and Forecast*, edited by Andrew Freris, pp. 71–96. Hong Kong: Salomon Brothers.

10. Ningpo Port Authority (1995). *Port of Ningpo* (in Chinese). Ningpo, June.

11. Organisation for Economic Cooperation and Development (1996). *OECD Economic Outlook*. Paris, June.

12. Planning Department and Port Development Board (1995). *Port Development Strategy Second Review: Executive Summary*. Hong Kong, October.

13. Port Development Board (1992). *Annual Report 1992*. Hong Kong.

14. _______ (1993). *Hong Kong Port Cargo Forecasts 1993/4*. Hong Kong, December.

15. _______ (1994). *Annual Report 1993/4*. Hong Kong.

16. _______ (1995). *Annual Report for the Period to September 1995*. Hong Kong.

17. _______ (1996a). *Hong Kong Port Cargo Forecasts 1995*. Hong Kong, February.

18. _______ (1996b). *Economic Contribution of Containerised Cargoes*. 7 June.

19. Rodwell, Simon (1989). *Boxes and Barnacles: The Story of Hong Kong International Terminals*. Hong Kong: Hong Kong International Terminals Ltd.

20. Sinclair, Kevin (1992). *The Quay Factor: Modern Terminals Limited and the Port of Hong Kong*. Hong Kong: Modern Terminals Ltd.

21. Taylor, D. A. (1991). *The Port of Hong Kong*. Hong Kong: Book Marketing Ltd.

22. Westlake, Michael (1991). "Port out of a Storm," *Far Eastern Economic Review* 18 July, pp. 56–57.

Index

About the Authors

Professor Leonard K. Cheng is Professor and Head of the Department of Economics at The Hong Kong University of Science and Technology. He studied economics at The Chinese University of Hong Kong (BSocSc 1975) and the University of California at Berkeley (MA 1977 and PhD 1980). Prior to joining HKUST in 1992 he was Associate Professor of Economics at the University of Florida, where he taught for 12 years. He was Visiting Professor at Fudan University in 1980 and has been Adjunct Professor of Shantou University since 1994. Professor Cheng's research interests are in international trade and investment, the theory of market structures, and the economics of technological innovation and imitation. He has published widely in leading economics journals, and currently is Associate Editor of the *Journal of International Economics*, and an editorial board member of the *Pacific Economic Review*. As part of his service to the community, he is a member of the Hong Kong Government's Economic Advisory Committee and a member of the Pacific Economic Cooperation Council's Trade Policy Forum Committee (Hong Kong).

Professor Yue-Chim Richard Wong is Professor in the School of Economics and Finance and Director of the APEC Study Centre at The University of Hong Kong. He studied economics at the University of Chicago (BA 1974, MA 1974, PhD 1981), and was Visiting Scholar at the National Opinion Research Center at the University of Chicago in 1985 and at the Hoover Institute of War, Revolution and Peace at Stanford University in 1989. His research interests are in political economy, macroeconomics, and housing markets. He has authored numerous books and articles and currently is Associate Editor of the Asian Economic Journal and an editorial board member of the Pacific Economic Review. In addition to his scientific work, Professor Wong is active in public policy research. He is founding Director of the Hong Kong Centre for Economic Research, an independent public policy research institution. He is a member of the Hong Kong Government's Economic Advisory Committee; the Pacific Economic Co-operation Council's Hong Kong Committee for Pacific Economic Co-operation; Hong Kong's Industry and Technology Development Council; the Finance Committee of Hong Kong's Housing Authority; and the University Grants Committee.

The Hong Kong Economic Policy Studies Series

Titles	*Authors*
Hong Kong and the Region	
❑ The Political Economy of Laissez Faire	Yue-Chim Richard WONG
❑ Hong Kong and South China: The Economic Synergy	Yun-Wing SUNG
❑ Inflation in Hong Kong	Alan K. F. SIU
❑ Trade and Investment: Mainland China, Hong Kong and Taiwan	K. C. FUNG
❑ Tourism and the Hong Kong Economy	Kai-Sun KWONG
❑ Economic Relations between Hong Kong and the Pacific Community	Yiu-Kwan FAN
Money and Finance	
❑ The Monetary System of Hong Kong	Y. F. LUK
❑ The Linked Exchange Rate System	Yum-Keung KWAN Francis T. LUI
❑ Hong Kong as an International Financial Centre: Evolution, Prospects and Policies	Y. C. JAO
❑ Corporate Governance: Listed Companies in Hong Kong	Richard Yan-Ki HO Stephen Y. L. CHEUNG
❑ Institutional Development of the Insurance Industry	Ben T. YU
Infrastructure and Industry	
❑ Technology and Industry	Kai-Sun KWONG
❑ Competition Policy and the Regulation of Business	Leonard K. CHENG Changqi WU
❑ Telecom Policy and the Rise of Digital Media	Milton L. MUELLER
❑ Port Facilities and Container Handling Services	Leonard K. CHENG Yue-Chim Richard WONG

Titles	*Authors*
❑ Efficient Transport Policy	Timothy D. HAU Stephen CHING
❑ Competition in Energy	Pun-Lee LAM
❑ Privatizing Water and Sewage Services	Pun-Lee LAM Yue-Cheong CHAN
Immigration and Human Resources	
❑ Labour Market in a Dynamic Economy	Wing SUEN William CHAN
❑ Immigration and the Economy of Hong Kong	Kit Chun LAM Pak Wai LIU
❑ Youth, Society and the Economy	Rosanna WONG Paul CHAN
Housing and Land	
❑ The Private Residential Market	Alan K. F. SIU
❑ On Privatizing Public Housing	Yue-Chim Richard WONG
❑ Housing Policy for the 21st Century: Homes for All	Rosanna WONG
❑ Financial and Property Markets: Interactions Between the Mainland and Hong Kong	Pui-King LAU
❑ Town Planning in Hong Kong: A Critical Review	Lawrence Wai-Chung LAI
Social Issues	
❑ Retirement Protection: A Plan for Hong Kong	Francis T. LUI
❑ Income Inequality and Economic Development	Hon-Kwong LUI
❑ Health Care Reform: An Economic Perspective	Lok-Sang HO
❑ Economics of Crime and Punishment	Siu Fai LEUNG

OTH030	Due Date	
7/25/97		